Animal Advocacy Documentaries in the Twenty-first Century

Animal Advocacy Documentaries in the Twenty-first Century is the first monograph dedicated exclusively to the study of this subgenre.

The book explores how the plasticity of nonfiction film engages with nonhuman otherness and with the rhetorical discourses of advocates and activists, while simultaneously reaffirming its claims to truth. It examines how documentary mechanisms associated with form and content are deployed by filmmakers to represent nonhuman animals as subjects deserving of moral consideration, effectively seeking an ethical response from viewers. Rhetorical devices at the service of the documentary genre and animal ethics discourses come together in the formulation of narratives and aesthetics that reflect filmmakers' awareness of the intertextual forces embodied by nonhuman animals in moving images.

Contributing to the fields of (Critical) Animal, Film and Media Studies by defining and historicizing these films, the book situates their identity in the context of Western mythographies, politics of sight, the systemic practices of the animal-industrial complex and contemporary self-reflexivity. It will appeal to academics, students, advocates, activists and people with an interest in nonfiction film, offering them an understanding of the historical and aesthetic place of animal advocacy documentaries, as well as a critical assessment of their limits and problematic interstices.

Claudia Alonso-Recarte is Associate Professor in the Department of English and German at the University of Valencia, Spain. She specializes in the field of (Critical) Animal Studies, with a particular interest in the representation of nonhuman animals in literature and film. She has published in leading academic journals, including *Gender, Place & Culture*, *MELUS*, *Men and Masculinities*, *Journal for Critical Animals Studies*, *Atlantis* and *Studies in Theatre and Performance*. She has served as a principal investigator of two funded research projects on animal advocacy documentary film and is the founder and director of the research group *Animals in Literary and Visual Cultures* (CULIVIAN) at the University of Valencia.

Routledge Advances in Film Studies

Place and the Moving Image
Essay Film Practice as Topography
Elisabeth Brun

Cinematic Starchitecture
The Celebrity Status of Urban Architectural Structures in Film
Edited by Merrill Schleier & Paul Newland

Posthumanism, Cognition and Cyborg Spectatorship
Amalgamated Cinema
Anna Batori

Narratives of Cultural Identity in Contemporary Arab Cinema
Voices from the Margins
Hania A.M. Nashef

India's Women Filmmakers in the Age of Netflix
Female Lead
Zeltzyn Rubi Sanchez Lozoya

River Space in Contemporary Chinese Cinema
Home, Locality, and Waterbodies
Mei Yang

Cinema and the Dream of the Living Machine
Joseph Zornado and Sara Reilly

Animal Advocacy Documentaries in the Twenty-first Century
Claudia Alonso-Recarte

For more information about this series, please visit: www.routledge.com/Routledge-Advances-in-Film-Studies/book-series/RAFS

Animal Advocacy Documentaries in the Twenty-first Century

Claudia Alonso-Recarte

Routledge
Taylor & Francis Group
LONDON AND NEW YORK

First published 2027
by Routledge
4 Park Square, Milton Park, Abingdon, Oxon OX14 4RN

and by Routledge
605 Third Avenue, New York, NY 10158

Routledge is an imprint of the Taylor & Francis Group, an informa business

British Library Cataloguing-in-Publication Data
A catalogue record for this book is available from the British Library

Library of Congress Cataloging-in-Publication Data
Names: Alonso-Recarte, Claudia author
Title: Animal advocacy documentaries in the twenty-first century / Claudia Alonso-Recarte.
Description: London ; New York : Routledge, 2027. | Series: Routledge advances in film studies | Includes bibliographical references and index.
Identifiers: LCCN 2026032340 (print) | LCCN 2026032341 (ebook) | ISBN 9781032661353 hardback | ISBN 9781032661384 paperback | ISBN 9781032661445 ebook
Subjects: LCSH: Documentary films—History and criticism | Animals in motion pictures | LCGFT: Film criticism
Classification: LCC PN1995.9.D6 A46 2027 (print) | LCC PN1995.9.D6 (ebook)
LC record available at https://lccn.loc.gov/2026032340
LC ebook record available at https://lccn.loc.gov/2026032341

ISBN: 978-1-032-66135-3 (hbk)
ISBN: 978-1-032-66138-4 (pbk)
ISBN: 978-1-032-66144-5 (ebk)

DOI: 10.4324/9781032661445

Typeset in Times New Roman
by Apex CoVantage, LLC

To Nacho, always.

Contents

Acknowledgments

This book is the result of the research undertaken as part of two consecutive projects funded by the Conselleria d'Educació, Cultura i Universitats (Generalitat Valenciana): "The Ethics and Aesthetics of Animal Advocacy Documentary Film in Twenty-first-century Western Cultures" (CIAICO/2023/046) and "Representations of Masculinities in Animal Advocacy Documentaries in English (2000–2021)" (CIGE/2021/100). Additional funding was granted by the Real Colegio Complutense at Harvard University, directed by Daniel Sánchez Mata. I am thankful to Prof. David Joselit, Prof. Sharon Harper and Ms. Denise Oberman, from the Department of Art, Film, and Visual Studies at Harvard for generously enabling me to conduct part of my research there.

I am extremely grateful to all the colleagues in the field of (Critical) Animal Studies who have enriched my knowledge and from whom I continue learning every day: Brett Mills, Randy Malamud, Claire Parkinson, Núria Almiron, Catia Faria, Laura Fernández, Olatz Aranceta-Reboredo, Nuria Máximo Bocanegra, and my dear friends María Teresa Lajoinie Domínguez and Margarita Carretero González. Special thanks go to one of my greatest role models, Jo-Anne McArthur, and to the people at We Animals, whose work is a source of inspiration to so many of us. Thank you as well to the students and young researchers whose courage and dedication ensure the continuation of research in (Critical) Animal Studies: Sofía Duarte, Olimpia Mailat, Betlem Pallardó, Esther Igual, and Michał Matuszewski. The completion of this book would not have been possible without the help of filmmaker Emilio Fonseca, who patiently and insightfully took the time to answer all of my very long emails with endless questions.

Thank you to my dear friends and fellow academics at the Department of English and German at the University of Valencia, who have supported me with warmth and kindness every step of the way: Ana Fernández-Caparrós, Sergio Maruenda, Carmina Gregori, Laura Monrós and Vicent Cucarella. Thank you as well to the friends I grow old together with, for holding my heart during grief and in times of joy: Mercedes González, Teresa Gómez, Patricia Martínez, David Yagüe, Manuel de la Fuente, Josep Maria Armengol, Óscar Porras, Ángel Diezma, Montserrat Borrego, Agustín Reyes, Amanda Bird, Begoña Ruiz, Jesús Moya and María de la Rica. I am immensely grateful to my sister, Paula, and her partner, Fiti, for their unconditional love and support and for always bringing a smile to my face. And to my in-laws, Herminia and Emilio, who from the beginning treated me as their daughter and have helped me find a place in the world.

To my companions throughout the years, for illuminating life with love and wonder, and for showing me your magnificent worlds: Turco, Tracker, Tosco, Tosca, Sito, Dusty, Maharajah, Cuentacondes, Taiga, Tundra, Ramsar, Odín, Nashoba, Shilajoe, Lekeitio, Timber, Tuca, Runt, Benny and Eleven, who briefly came to us in the final stages of writing this book before returning home.

Kindness and compassion toward nonhuman animals are lessons that I would never have learned without the guiding hand of my parents, Enrique and Ana, who initiated me into the journey of experiencing shared vulnerabilities and love across species. My father's work ethic and dedication to the field of animal law and my mother's care for all the companions who shared their lives with us are lessons that I will always treasure. I love you, and I hope I make you proud. Mom, I miss you more than words can say.

My most heartfelt gratitude goes to the people who are my home and who can call these pages their own. To my beautiful, smart, mischievous, Frankie, our companion for the past thirteen years, who has been by my side during the writing of each and every word in this book and who now, at the age of almost 17, brings with her the puncturing reminder of the fleetingness of shared, blissful time. To my partner, Nacho, who from the beginning has encouraged and believed in my work, helping me navigate through exhaustion and doubt. Thank you for always being so present, for understanding the path, and supporting me in every aspect of life. Always my soulmate, always my team. And lastly, thanks to our incredible children, Mina and Kike, who have made life so much more purposeful and meaningful than I could have ever anticipated. You have effortlessly created a world of joy, compassion, and laughter where I wish I could freeze time so as to never know the pain of nostalgia. You are my anchor.

Introduction

Animal Advocacy Documentary as a Subgenre

In recent decades, the proliferation of animal advocacy documentaries has attracted the attention of audiences, scholars and cultural critics. This book provides a comprehensive discussion of the paradigmatic features shaping these types of films and enabling their consideration as a legitimate subgenre[1] in the twenty-first century. It focuses on issues involving representation (of nonhuman animals, advocates and activists), focalization and framing, form (documentary modes, cinematography and editing), argumentative rationale and the aesthetics of rhetoric.[2] It also grapples with the complexities arising from interdisciplinary approaches to identity politics and from epistemologies that center nonhuman animals as sentient beings and moral subjects. In their attempt to anchor nonhuman animals out of the marginal, these films have deployed a series of identifiable strategies that depend on the plasticity of nonfiction and the moving image and on their ability to convey alternative narratives of "truth." Together, they have fabricated discourses that speak to a broader intertext of animal rights media, but that are also able to retain their singularity as a body of works that consciously and self-reflexively explores its own potential. Over twenty years ago, Jonathan Burt observed how there was very little written about the history of animal rights films, their emergence and the role of the Internet in spreading their hallmark type of imagery (2005, 347). Very important scholarly work has been done since then to fill this void, and this book aims to become part of the conversation by looking at these films holistically, as exemplars and practices of a subgenre, not as scattered or isolated forms of storytelling.

A Subgenre Is Born

Animal advocacy documentaries of the twenty-first century are heirs to the shocking films, videos and footage that, beginning in the 1980s, became effective tools used by the animal rights and liberation movement to raise awareness and denounce hidden practices in animal exploitation industries. Scholars and activists launched and diversified the movement in the last quarter of the twentieth century, delivering a highly articulated network of theories where philosophies of welfare, deontology, utilitarianism, abolitionism, contractualism and care ethics questioned and condemned humans' systematic and institutionalized exploitation of other species as part of what has come to be known as "the animal-industrial complex."

DOI: 10.4324/9781032661445-1

Barbara Noske coined this term in 1989 to refer to the systemic production and consumption of nonhuman animals activated and sustained by capitalist forces seeking expansive markets and profitability. The animal-industrial complex commodifies nonhuman animals, manufacturing them as raw materials and normalizing (and obscuring) the massive violence inflicted upon their bodies and minds. Working definitions of the term have continued since then (Twine 2012; Hunnicut et al. 2025), with the aim of conceptualizing it as a framework that not only is descriptive of the corporate and political power relations with stakes in the sectors and industries involved but also serves to assess the magnitude of the complex and "to theorize transition away from it" (Hunnicut et al. 2025, 3). From the beginning, documentaries have substantially contributed to the exposure of the mechanisms of the animal-industrial complex and the layers of discourses concealing the capitalist interests it protects. Just as importantly, they have collectively devised formal and discursive strategies that are grounded in the conventions of activist nonfiction films but that have evolved into audiovisual grammars and semiotics that are identifiably anti-speciesist. In other words, the ongoing production of documentaries has laid the foundation for their consideration as a subgenre of films that have forged, extended and refined distinct narratives, indexes and rhetorics, weaponizing them against the animal-industrial complex.

The films and videos that activist organizations and advocates produced and disseminated in the 1980s set the tone for the development of the subgenre. Titles such as *The Animals Film* (Schonfeld and Alaux 1981), *Unnecessary Fuss* (Newkirk and Pacheco 1984), *Britches* (Gruen 1986) and *Hidden Crimes* (Burgos 1986) formed the first body of audiovisual works to graphically expose the normalized, systemic cruelty and abuse of nonhuman animals. During this period, vivisection and laboratories were especially targeted, following in the footsteps of the nineteenth-century British and American humane movements. The impact of some of the films became a clear sign of their potential to reveal hidden practices and pressure for reform: *Unnecessary Fuss*, with a duration of less than thirty minutes, was the result of People for the Ethical Treatment of Animals' (PETA) editing of more than sixty hours of footage recorded by researchers at the University of Pennsylvania's Head Injury Clinic. The footage, which showed clear evidence of researchers' cruelty and negligence toward baboons, was stolen by the Animal Liberation Front (ALF) and handed to PETA. The film was distributed to media outlets and Congress, and led to the temporary suspension of the clinic's grants. More importantly, the public indignation it caused was an indication of mounting historical momentum for possible change. In a similar line, in 1989 a woman by the name of Beck Sandstedt recorded over forty hours of the miserable conditions of "downers" (the weak, injured and moribund livestock that reach the stockyard) in St. Paul, Minnesota. The evidence caused United Stockyards to change their policies so that it would no longer be profitable for livestock owners to deliver downers (Finsen and Finsen 1994, 1–2). The recordings were edited by Farm Sanctuary into an eighteen-minute video titled *The Down Side of Livestock Marketing* (1991). Other films such as Tom Regan's

We Are All Noah (1986) and *Voices I Have Heard* (1988) created a platform for the exposition of animal ethics arguments in relation, respectively, to religious faith and activist trajectories. It soon became evident to activists and advocates that collecting footage was only the first step, and that much effort had to be put into strategizing about how to effectively bring these images before the public (Phelps 2007, 292–98). Short and feature-length documentaries and videos became a fundamental instrument that, if well distributed, could serve the animal rights movement well.

These early documentaries initiated a formal aesthetic which later filmmakers would use to negotiate their own style and voice. The exploration of alternative rhetorical angles – by way of cinematography, editing, intertextuality, management and delivery of terms and arguments – is not just caused by the endogenous, immanent growth of aesthetic genres toward organic, self-reflexive change. External forces are also at play, especially when considering a documentary subgenre whose very reason for being (its *telos*) is based on the premise that film can precipitate reform and inspire transformative change in the real world. The zeitgeist and sociopolitical climate drastically affect the direction of these changes and motivate the stylization of filmic proposals. Consequently, there are a couple of assumptions regarding twenty-first-century animal advocacy documentaries that mark a transition from the material produced in the 1980s and 1990s: First, because of the impact of the movement, viewers (by which I refer to average mass media consumers in Western societies) have at least some knowledge about there being such a thing as an animal rights debate, and that such a debate concerns everyday practices having to do with human consumption of food, medication, clothing and entertainment. That is, they have been exposed to the existence of moral positions that are strongly critical of the animal-industrial complex. Second, viewers have acquired perspectives resulting from exposure to mass-media narratives about climate change, species extinction, fake news, post-9/11 anti-terrorism and nationalism, gender and racial politics, and celebrity culture, all of which greatly determine and influence the shaping of advocacy discourses. The shifts in meaning making and storytelling on the part of twenty-first-century media, which acts as a thermometer of sociopolitical tensions and, simultaneously, as a prescriptive force, have of course also affected theoretical and practical approaches to the documentary genre. As a result, in the last twenty-five years, academics and filmmakers have largely explored the significance of signature issues such as the claims to truth, authenticity and authority in documentary praxis.

The documentary genre cloaks itself under claims of objectivity – a pretension that has always sparked criticism (Chapman 2009, 48–71; Corner 1996) and pushed the genre into exploring the plasticity of its forms of storytelling. In exercising their relation to reality, documentaries do not just "inform": they can entertain, authenticate, distort, (de)mystify, hyperbolize, belittle, exploit and rhetorize their subject matter in multiple ways. The apparatus of narratives, articulated through cinematography, montage and editing techniques, implodes the possibility of objectivity, a force against which the documentary inertially competes at the same time, for at its heart is the claim to truth, or at least to some semblance of it. Documentaries are producers of knowledge determined not only by their content and form but also by the way they are circulated and experienced in the world that manufactures them (Corner 1996).

Activist documentaries in general constitute a "radical interrogation and alternative perspective" (Corner 2002, 260) on the dominant versions and discourses that serve to justify institutional structuring, social order and policymaking. "Implicitly, sometimes explicitly, the documentary discourse attempts a criticism and a correction of other accounts in circulation" (Corner 2002, 260). Activist documentaries' claims to truth are fundamentally connected to affective responses. Feeling that the state of affairs is wrong, cruel or unjust may be prompted by shocking images, carefully reasoned arguments, focalizations from the point of view of marginal subjectivities, or cyclical repetitions of motifs and patterns that elicit empathy and compassion toward the "other." As acts of civic, democratic participation, activist documentaries accentuate the perspectives of organizations and advocates who occupy positions of aesthetic and/or moral guidance but generally lack actual governmental or decision-making power (Aguayo 2019). Whatever the subject matter, the activist documentary ensures that it is framed as part of a conflict where political, economic or sociocultural interests are at stake. It is important for these films to make it clear for the viewer that such conflict does exist, and that it is a sign of a grander scheme of power relations that seeks to quell, conceal or dismiss the situation as accidental or residual. These films produce knowledge with their own representations of truth, hoping to mobilize viewers to take action. In the case of animal activism, action can range from boycotting companies and products, signing petitions or adopting vegan lifestyles, to actively protesting against different administrations or companies, supporting animal rights or welfare organizations, or educating people (e.g., by screening the films for new audiences). Lighting a spark in the conscience of the viewer can, in the long run, lead to the consolidation of anti-speciesist initiatives.

Animal advocacy documentaries are deserving of critical attention because of the notable increase in their production in the last twenty-five years, and because they greatly contribute to the body of media that seeks to raise awareness about the animal-industrial complex. As part of this audiovisual media that includes other types of films, news reports, publicity and online videos and clips, they should be regarded not as isolated material, but as historically determined productions that are consciously shaped into discourses, narratives and other forms of persuasion that speak to a wider intertext of activism. Quantitatively speaking, the sheer amount of activist media available online through social networks, websites and streaming platforms is overwhelming and seemingly boundless. However, the repetition, extension and refinement of certain narratives, strategies and structures allow us to apprehend and classify these media discourses across transversal dimensions of recognizable patterns. Animal advocacy feature-length documentaries constitute an important portion of this media and foster what Angela Aguayo (2019) refers to as "documentary commons," that is, a kind of public domain in which the production, circulation and engagement with films is exercised as a practice of participatory, democratic exchange shaping identities and communities in a social order. In this sense, animal advocacy documentaries are experienced as part of a public realm of interaction, as opposed to functioning solely within the boundaries of the cinematic institution.

The very public visibility of (collective) viewer response is proof of the communicational dynamics of the documentary commons. Audiences manifest tacit agreements built on shared tastes and expectations which are often associated with the prospects anticipated by the (sub)genre. Viewers can, in fact, become very protective of a subgenre: when the first season of the sensationalist Netflix docuseries *Tiger King* (Goode and Chaiklin 2020) drew comparisons with *Blackfish* (Cowperthwaite 2013), critics immediately expressed their disapproval (Nuwer 2020; Trumbore 2020). The anti-captivity message of *Blackfish* had drastically damaged the reputation of SeaWorld, leading to changes in policies and in their use of orcas, and had raised public awareness about the abuse of cetaceans. *Tiger King*, on the other hand, operated between true crime and reality television formulas, and largely focused on the bizarre antics of American big cat owners, zookeepers and rescuers, massively attracting binging audiences. The mere comparison between the two productions seemed outlandish and showed the extent to which both filmmakers and critics were aware of how public perceptions about types of subgenres could be manipulated for the purpose of publicity.

As suggested earlier, the transversality of recognizable patterns does not constrain animal advocacy documentaries to a formal or thematic paralysis. The films' engagement with intertextuality leads to the manifestation of variations in such patterns, inviting viewers to regard them as a microcosm of representative characteristics. This representativeness is both disseminated and prescribed by the instruments of canon formation, such as (social) media, online platforms, movie theaters, academic studies, reviews and film festivals (Vallejo 2020) – or even this very book. Acknowledging the existence of a canon in animal advocacy documentary film does not just mean validating a selection of works as exemplars of a broader corpus; it also legitimizes the subgenre and lends visibility to advocates as a community united by shared sensibilities. Advocacy and activism instrumentalize the canon – it is not meant to function as a mere object of aesthetic contemplation, but to be effectively wielded against perceived injustices.

The status of these films as cultural products must also be taken into account. While visibility and the amount of viewership are factors with a definite impact on canon formation, the production and dissemination of these films reveal different manufacturing strategies. Some films receive private and/or public funding for their production and/or distribution, which more effectively assures their exposure in more mainstream, commercial circuits, such as television and movie theaters. The films released in movie theaters help place the subgenre in the production chain of cinema; they become part of the social rituals and practices of motion picture distribution that condense aesthetics, ideology and spectatorship in spaces and mediums designed to reach mass audiences. This form of distribution raises their chances of marketability and critical recognition, to the point where some of these films have been Academy Award winners (*The Cove* [Psihoyos 2009], *My Octopus Teacher* [Ehrlich and Reed 2020]), nominees (*All That Breathes* [Sen 2022]), shortlisted finalists (*Project Nim* [Marsh 2011]), or British Academy of Film and Television Arts (BAFTA) nominees (*Project Nim*, *Blackfish*, *Cow* [Arnold 2021]). Streaming services provided by platforms such as Netflix, HBO or Amazon Prime (which have also contributed to original productions) have also increased viewership and helped to spotlight some of these films. Other films are independently produced, or are made by smaller, specialized production companies, or depend on private donations and fundraising for their completion, exemplifying how the production process itself can materialize as a communal form of advocacy and activism. Although these films are not easily accommodated within mainstream cinematic or television culture, their (limited) screening in smaller, less commercial movie theaters, and at activist and academic events, provides them with a level of exposure that can steadily grow because of word of mouth. Many of them also become available on platforms such as YouTube, Vimeo and Watch Documentaries, fomenting their visibility. Some fare well in terms of popularity and viewer feedback, while many others do not achieve significant distribution or prompt widespread response (although they remain indispensable for keeping the subgenre alive). The films discussed in this book have been selected because of how they contribute to the representativeness of the animal advocacy subgenre, and because of how they reflect and influence its evolution. Whether they qualify as more mainstream, commercially successful types of productions or are more aligned with the processes and aesthetics of independent filmmaking is a factor that will be considered to further critical commentary, but not as a criterion for their inclusion. Therefore, this book acknowledges its scholarly interest in the more canonical, feature-length documentaries because of their representativeness, but it also occasionally draws on shorter and lesser-known films to strengthen its arguments.

Nonhuman Animals and the Media

We generally do not think of being in regular contact with nonhuman animals unless we are pet owners or it is part of our profession. And yet we routinely consume nonhuman animals in our daily affairs without giving it a second thought. We ingest them, we wear them, we verbally appropriate them through common metaphors in formal and informal registers and we are visual recipients of their omnipresence in mass media. Through them, we have cognitively practiced the dissociation of metonymies: to us they are long-lost wholes, forgotten totalities that are represented by pieces. Humans transform nonhuman animals into absent referents (Adams 1990) and linguistic constructs (Dunayer 1995). They have become very practical for our anthropocentric affairs, as Claude Lévi-Strauss's somewhat clichéd claim that "natural species" are "good to think" (1964, 89) summarized. The pieces of them that we systematically consume have conveniently ceased to signify the completeness of their individuality.

The widespread presence of real nonhuman animals on the twenty-first-century screen cannot, as of today, escape a myriad of ethical dilemmas concerning their exploitation in the entertainment industry, how this violates their integrity and privacy, and how they are objectified to serve anthropocentric narratives. With capitalism as the prime suspect behind the aforementioned dissociation between part and whole, several scholars (Baker [1993] 2001; Rothfels 2002; Nibert 2013; Shukin 2009; Molloy 2011; Almiron et al. 2016; Mills 2017) have provided invaluable insights into the deconstruction of the representation of nonhuman animals in postmodern popular culture, culminating in the emergence of a new field in its own right: Critical Animal and Media Studies (Almiron et al. 2016). These studies converge on the understanding that the capitalistic way of relating to the "other" is mediated by a form of exploitative consumerism that provides material and financial profit for those in power – this is the essence of the animal-industrial complex. And to effectively perpetuate such systems of oppression, the media assists in spreading the idea "that capitalism, if not the ideal system, is at least the best economic and social system humans can develop" (Nibert 2002, 15). Mass media disseminates narratives that sanitize the image of capitalism, representing the oppressed in ways that offer little (or what is deemed to be proportionally reasonable) challenge to the status quo. Sometimes, the challenge itself becomes absorbed and assimilated by ideology in a display of the elite classes' institutional power.

The media, as the social institution that it is, is a prime agent in the establishment of what is morally acceptable because it creates and spreads an understanding of the qualities that make good citizenship across different socioeconomic units and contexts (the family, the workplace, public spaces, the larger body politic etc.). The media describes and prescribes moral values, validating or condemning practices and beliefs.

How we frame nonhuman animals, how we dramatize and narrativize them in the name of anthropocentrism, is no passive affair. There are corporate interests underlying mass media representations of them; interests that elusively wrap themselves in a confusing rhetoric that excuses and perpetuates exploitation. When not on the grounds of tradition and national sentiment, speciesism is justified in the media by a pseudoscientific claim that humans are omnivores (a habit which we cannot seem to shed), by the alleged necessity of using nonhuman animals as research and test subjects, or by blurry and imprecise definitions of conservation and environmentalism that warrant the existence of zoos and aquaria. Media representations of nonhuman animals constitute a fundamental way through which humans manufacture the consent to their exploitation and abuse (Almiron 2016, 17).

Consider the following case as a common example of Western media's framing of nonhuman animals: on April 10, 2022, the weekend newscast of Telecinco, an FTA Spanish channel, featured several stories involving nonhuman animals. Beginning its almost 34-minute run with hard political news (the presidential elections in France, the war in Ukraine, internal affairs and brief notes on foreign protests), there then comes the story of the continued lockdown of 25 million people in Shanghai. The story focuses on the extenuating circumstances of a population on the verge of chaos and shows images of sacks full of cats and dogs piling the streets – a landscape explained by a detached voice-over informing that pets belonging to COVID patients will be destroyed. In an uncanny twist, the story ends with images of a robot dog patrolling the empty streets with a loudspeaker reportedly asking citizens to stay indoors. The robot dog is referred to by the voice-over as an "artifact."[3] The following story is about the discovery of a 50,000-square-meter warehouse in a Valencian town where approximately 29 million euros worth of illegal big game trophies are stored. The voice-over clarifies that the suspect involved in the trafficking is the son of a well-known businessman. The next story marks the transition to soft news, spotlighting the traditional "pesca del Campanu" in Asturias, that is, the fishing of the first salmon of the season. This year's first fish, weighing approximately 7 kg, has been auctioned at 13,200 euros. "A fine specimen," an enthused reporter says, that continues a sixteenth-century tradition. This is then followed by a story about the reinvention of another long-standing practice: falconry. As animal-assisted therapy is becoming increasingly popular, the news anchor explains, new formats have emerged, such as therapy with birds of prey. The story follows a falconer "at work" in an old people's home, allegedly helping them exercise "good emotions," attention, memory, balance and concentration. How he apparently goes about this, the images show, is by having an owl and an eagle fly close to the residents and letting them pet and stroke the birds. "A millenary profession, falconry, at the service of the most vulnerable," the voice-over praises. Toward the end of the newscast, viewers are also informed of the opening of the bullfighting season on Palm Sunday in the Plaza de Toros de Las Ventas, Madrid, the "temple of bullfighting." Emilio de Justo is to face six bulls before 20,000 spectators in what promises to be a thrilling and "glorious" afternoon.

This Telecinco newscast condenses systematized forms of representation of nonhuman animals that are consistent with speciesist framings and narrativizations. In each and every story, nonhuman animals are victims of human activity. They pay with their lives (or their freedom) for their coexistence with humans in different settings, and yet the verbal and visual discourses creating the stories to which they are central do not acknowledge their individuality, their vulnerability or their victimhood. Whether treated as expendables of the pandemic, impressively valuable illegal pieces, or the inconsequential pawns of national traditions, the possibility to document their abuse is discarded, as the conventions of hard and soft news carry on without them. John Berger famously described how in the context of modernity,

> animals are always the observed. The fact that they can observe us has lost all significance. They are the objects of our ever-extending knowledge. What we know about them is an index of our power, and thus an index of what separates us from them. The more we know, the further away they are.
>
> ([1980] 2009, 27)

This disappearance of the nonhuman animal in modernity is a form of biopolitical rendering that restricts human contact with other species to the consumption of pieces and refractions of who and what they really are. As Steve Baker stated in his groundbreaking *Representing Animals*, "there [is] no unmediated access to the 'real' animal" ([1993] 2001, xvi) because they can only be approached through representation. Mass media functionally precludes their completeness as subjects and their ability to respond – and our ability to respond to them. Not unlike Max Horkheimer and Theodor Adorno's ([1947] 2002, 94–136) theory about the capitalist commodification of art through the culture industry to make it profitable, nonhuman animals have become absorbed by mass media and technology in a way that reinforces their designated state as consumables. As Akira Mizuta Lippit contends, technological media perpetuates a state of undeadness of nonhuman animals that confronts their historical vanishment, as they are left to inhabit "the cryogenic topographies of technological reproduction" (2000, 161).

Furthermore, as Claire Molloy suggests, nonhuman animals in the media are objectified in such a way that they especially align with the expectations and affects associated with soft news. They are "easily constructed as characters in dramas, conflicts and romances" (Molloy 2011, 7), satisfying our assumptions about conventionalized genres. However, Molloy also reminds us that "not all news media representations trivialize animals. Public interest in conservation, animal welfare, animal rights, ethics and the environment has prompted the inclusion of animal stories that blur the line between soft and hard news" (2011, 7). These types of approaches are perhaps symptomatic of the "animal turn" (Merskin 2016) in the media, which interrogates the disseminated representations and discourses surrounding nonhuman others, paving the way for the emergence of the field of Critical Animal and Media Studies. Animal advocacy documentaries are an important part of the arsenal of the "animal turn": they form a body of narratives that in essence attempt to de-objectify nonhuman animals, to represent them as sentient beings of moral worth in and of themselves, not as instruments for anthropocentric purposes.

Countercultural Representation

One could argue that animal advocacy documentaries still constitute forms of representation (which of course is accurate, as there can be no divorce between the media and representation) in which nonhuman animals are intentionally and selectively depicted as victims, thus feeding into populism and propaganda. The counterargument is that this form of representation is not just more truthful: it is the *only* truth. That nonhuman animals are conscious, sentient beings that are self-aware and can experience a wide range of emotions is established as factual and undeniable. That humans systematically exploit them is established as factual and undeniable. As Carrie Freeman and Scott Tulloch state in their study of a sample of animal liberation documentaries, "even when documentaries avoid direct comparisons to humans, all seek to convince viewers that other animals have consciousness and sensitivity – in essence, agency" (2013, 119). There is little wiggle room allowed by animal advocacy documentaries in this regard, and having grounded the truth in such claims, nonhuman animals can only be represented in ways that do not deny or downplay their victimhood. As Patricia Aufderheide observes regarding activist documentaries in general, "they, like propaganda films, deserve attention from anyone who wants to understand the techniques of persuasion – and nothing persuades like reality" (2007, 78). Without compromising their convictions about the subjectivity and victimhood of nonhuman animals, different films can explore thematic and formal variations that project alternative styles and aesthetics within the subgenre of animal advocacy documentaries. What remains nonnegotiable and constant, uniting them as a subgenre, is the who-ness of the nonhuman animal subjects and their status as victims. It is important, therefore, to point out that animal advocacy documentaries are not equivalent to what Anat Pick (2018) terms "vegan cinema," which proposes an aesthetic sensibility derived from the analogy between the consumption of the gaze – driving the cinematic apparatus – and the act of devouring. For Pick, the schema of veganism foregrounds an aesthetic possibility of non-consumption, creating a "*non-devouring gaze*" (2018, 127; emphasis in original), and "indicat[ing] the ways in which films, or moments in films, call attention to the complicity between the desire to look and to eat, and in so doing invoke a vegan way of being in the world" (2018, 128). Vegan cinema is a sensibility, applicable to fictional and nonfictional genres, and it "does not refer to films about or in favour of veganism. Its rhetorical mode is not one of agitation or advocacy" (2018, 127).

To deny the artificiality of representation in documentaries is both naive and counterproductive. Even without the support of academic studies on documentary film (which have emphasized the unnaturalness of the genre as a construct and discourse), today's average citizens are probably sufficiently media-savvy to intuit that documenting, informing and reporting add up to forms of storytelling where an indefinite number of choices are made by filmmakers to lead viewers to particular meanings and values. Bill Nichols distinguishes between the notions of reproduction and representation: reproduction is judged on the grounds of fidelity to an original; representation is judged "by the nature of the pleasure it offers, the value of the insight or knowledge it provides, and the quality of the orientation or disposition, tone or perspective it instills," he writes. "We ask more of a representation than we do of a reproduction" (Nichols 2001, 20–21). That documentaries are nonfiction films – that is, they address historically grounded aspects of the world as we know it (or as we think we know it) – does not mean they are reproductions. They inescapably explore representation using image, sound, structure, order, eloquence and persuasion, balancing their commitment to reality as they strive to validate and authenticate the portrayal of the world as depicted in the film. Documentaries qualify as what Nichols calls "discourses of sobriety" because they "[claim] to address the historical world and to possess the capacity to intervene by shaping how we regard it." They are determined "to make a difference in how we regard the world and proceed within it" (Nichols 2001, 39). Viewers engage in a tacit pact with the filmmaker's representation, expecting the film to somehow inform, refine, reinforce, challenge or in some way influence their worldview.

It is also hard to establish a clear boundary between what is commonly referred to as a documentary and other nonfictional audiovisual media. The stories of the Telecinco newscast, a YouTube or TikTok video, reality television etc. are formats that also advance their own claims to truth. They may not answer to our commonly shared view of what properly qualifies as a documentary, but they certainly reproduce many of its rhetorical features. Jihoon Kim's (2022) recent study of twenty-first-century digital innovations (including how images are documented, circulated and displayed in different platforms) challenges the stability of the documentary genre as a signifier and points to the versatility of its aesthetic, its rhetoric and its epistemology. The reasonable conclusion to these complications is that, as independent filmmaker Jill Godmilow once stated in an interview, the term "documentary" is convenient but hopelessly limited, having a reductive effect on the possibilities of what nonfiction can do (1997, 80–81).

As a user-friendly concept, the notion of documentary is further problematized by its ramifications into subgenres that can be classified by subject matter (what it deals with) or form (how it deals with it). As suggested earlier, animal advocacy documentaries mostly fall under the label of the activist documentary, which denounces some practice or situation that is perceived by the filmmaker (and/or the sponsoring organization) as unjust or morally questionable, demanding public attention and accountability. The activist documentary presents a conflict with its antagonistic sides, usually pulling viewers into moral dualisms of good versus bad, the exemplary versus the shameful, the righteous versus the greedy. These antagonisms guide viewers toward anagnorisis and offer as a solution to the crisis a limited number of choices. The more affirmatively "activist" a film, the smaller the space for ambivalent response: the rhetoric of persuasion will be more explicit, encouraging viewers to follow what has been portrayed as the more ethical path. The dichotomy is additionally emphasized by formal conventions that viewers have come to identify with the subgenre as authenticators of truth. For example, Freeman and Tulloch note the common use of "grainy, low-resolution images that typify undercover footage captured on concealed handheld cameras," as activists trespass or infiltrate themselves into factory and fur farms, laboratories, slaughterhouses, breeding farms, mills or facilities. "However," they go on to add, "this quality, or lack thereof, does not detract from the films. As a semiotic encoding of authenticity, the low-tech style enhances their effect, serving as an unambiguous cue to viewers that the footage had to be captured undercover" (2013, 114).

Karen Whedbee states that the activist genre "functions as a platform for exposing social problems and expressing perspectives that would otherwise be ignored by mainstream media" (2019, 30). These perspectives, I would add, do not necessarily compete against missing information, but against the misinformation and disinformation that is spread by mainstream media regarding the conflict. In other words, an activist filmmaker may not just be shedding light on an issue unbeknownst to the public (i.e., the media does not provide a dominant narrative about it, usually because of a lack of interest); they may also frequently shed light on an issue about which there *is* a fairly consolidated mainstream narrative, which for different reasons is deemed by the filmmaker as wrongful and biased due to corporate, political or social interests. In the case of animal advocacy documentaries, there is an inherent struggle to make something of the aforementioned "undeadness" that is characteristic of nonhuman animals in visual media – to represent them meaningfully as subjects worthy of moral consideration. Animal advocacy narratives compete against well-manufactured, consolidated mainstream images intended to relieve social conscience. These images are connected to beliefs, such as the idea that nonhuman animals kept in farms are happy suppliers of their eggs, milk and flesh for human consumption; that dog breeds are a genetic guarantee of temperamental traits that justify their stigmatization (think of the pit bull scare in recent years); that in laboratories suffering is properly minimized or nonexistent; that cetaceans and pinnipeds are consensual participants happy to splash around in their tanks; or that endangered species are plentiful (their technological reproduction conveys this false idea of abundance). These mainstream images that mechanically sustain the manufactured consent to animal exploitation are naturalized into social and cultural practices that prescribe discourses on normalcy.

At the same time, the increasingly diversified methods providing access to marginalized information and points of view should not be underestimated. Animal advocacy documentaries compete against the mainstream mediatization of nonhuman animals, but perhaps today it would be a bit of an overstatement to claim that the average media consumer is completely blindsided by these representations. As I mentioned early in this introduction, the animal rights movement of the last quarter of the twentieth century did an excellent job in raising awareness about the animal-industrial complex (at least in the United States and in Britain, where the movement was mostly active). Such was their success in the media (especially in the 1980s) that animal rights protesting soon became a staple of – forgive the apparent oxymoron – mainstream counterculture, by which I mean the mainstream's yielding to some controllable level of dissent. The image of a healthy democracy must be carefully protected by satisfying representations of freedom of speech, freedom of thought and freedom of conscience. As Amitai Etzioni already noted about the United States in 1970, demonstrations "seem to be becoming a significant mode of political expression and a conventional, almost integral part of our democratic process" (1970, 8). Images of a controllable counterculture are integrated into the dominant narrative as helpful evidence to show that some level of dissidence (milder forms of protesting, demonstrations and civil disobedience) is tolerable and even encouraged by democracy, thus pacifying citizens' anxieties about state monitoring and surveillance and legitimizing the democratic model. Of course, in mainstream media there are limits to what is deemed an "acceptable" counterculture and what is condemned as radicalism and extremism. But, for now, suffice it to point out that the assimilation of animal rights issues into popular culture in the late twentieth century and its standardized coverage ever since have informed the average mass media consumer enough to intuit that there *is* an alternative moral perspective on animal exploitation, and that an individual has both the power and the responsibility to learn more. Target viewers' previous knowledge is considered when filmmakers embark on a project. John Duvall writes that

> [f]ilmmaking takes place within a particular social context, assuming a certain level of awareness and congruence of values within its audience . . .. [A] filmmaker frames her approach to a topic based on the audience's anticipated knowledge of and a predisposition toward her topic.
>
> (2017, 9)

The resulting choices made by filmmakers, which mark the leading directions explored by animal advocacy narratives, are the focus of this study, as they articulate a rhetoric that seeks to shake the conscience of spectators and remind them of their power to take some form of action, however small.

Competing Documentary Subgenres

The premise of this book is that animal advocacy documentaries have an identity of their own meriting consideration as a subgenre equipped with its own canonical works. This task requires not just tracing transversal tropes across form and content and assessing how they are creatively reinvented or subverted by filmmakers but also establishing where the boundaries of the subgenre are drawn (or, in some cases, blurred). Because of their thematic concern with nonhuman animals, their distinction from the subgenre of wildlife films seems a reasonable place to start. Derived from the exoticism of earlier ethnographic film and big game hunting reels, wildlife documentaries tend to produce representations of nonhuman animals in their "natural habitat," where they must compete to survive. How these representations deal with mode (the tragic, the comic, the melodramatic) is strongly bound to purposeful interpretations of ethological findings. They, in fact, constitute perfect examples of how nonhuman animals are patterned out as palimpsests in which to inscribe humans' cultural ethos and tropes. In contrast with the animal advocacy documentary, in which the authentication of the victimhood of nonhuman animals rests on their representation as sentient beings, traditional wildlife documentaries have long exploited nonhuman animal bodies and behaviors as semiotic catalyzers with an intrinsic potential to entertain, educate and indoctrinate through the powers of anthropomorphism. As Randy Malamud insightfully contends, "cultural frippery" hinders our sight of authentic nonhuman animals, perhaps because these "would be too depressing, or too scary, or too boring, for the viewer to endure" (2012, 72).

The wildlife documentary has claimed a solid position in the history and historiography of documentary filmmaking, and its versatility to adapt to audience demands has allowed it to systematically experiment with form. These films exemplify how historically ingrained voyeurism of other species can be capitalized on. Watching nonhuman animals has proven to be quite a lucrative activity in Western cultures, leading to the perpetual romanticizing of the rugged, self-made frontiersman (prepped today in outdoorsy outfits), fun family outings to national parks (with the promise of catching a glimpse of some rare species), zoos and aquaria (where sightings are guaranteed), or the multifarious forms that the touristic pursuit of wildlife can take (from safaris to whale watching). Not unlike the successful exploitation of remote non-Western human cultures that was popularized more than 100 years ago with Robert Flaherty's *Nanook of the North* (1922), the documentary's reaffirmation of our status as the powerful watchers objectifies the watched beings. Furthermore, the fetishistic pleasure aroused by gazing at wildlife (especially the more charismatic fauna such as large carnivores, cetaceans, or birds of prey) reinforces our role as consumers when technology relocates the nonhuman animal presence onto screens. Despite the frequent recital of the need for conservation and the threat of extinction, screens make wildlife seemingly abundant, a cornucopia of species populating the planet.

Wildlife documentaries also parade a range of behaviors that could be staged or might not even constitute a representative part of what those individuals spend most of their time doing, yet they conveniently surrender these behaviors to anthropocentric interests. "Even the animals themselves do not experience their lives as they are portrayed on film and television," Derek Bousé noted in his classical study of the genre. "Condensed into an hour or less, with the boring bits cut out, their lives appear far more busy and event-filled than they really are or ever could be" (2000, 8). Canonical and blue-chip wildlife and natural history films, from the mid-twentieth century onwards, set the tone and standard patterns that the narratives would take. Often credited is Disney's *True-Life Adventures* series, which in the 1950s crafted wildlife narratives that were "sentimental, anthropomorphizing and steeped in postwar ideologies of progress and individualism, homeland prosperity, and so-called family values" (Chris 2006, 28). These undertones were maintained for decades, perhaps culminating with Luc Jacquet's Academy Award-winning *La Marche de l'empereur* [*March of the Penguins*] (2005), which delivered such an excessively romanticized version of emperor penguin breeding behavior that actor and director Bob Saget parodied the film with *Farce of the Penguins* (2006). Broadcasting corporations such as the BBC have for decades released series after series of a more "scientifically sound" study of nature and wildlife. Some production companies soon found that by branding their films with the face and expertise of a white male authority they acquired the necessary touch to reinforce their claims at objectivity at the same time that they reasserted national sentiment (with figures such as David Attenborough in the UK, Félix Rodríguez de la Fuente in Spain, and Jacques Cousteau in France).

But when ideologies and social values are used to describe and prescribe nonhuman animal behaviors that are conveniently edited to follow a narrative pattern, ethical dilemmas arise. Brett Mills's studies on wildlife films' violation of nonhuman animals' right to privacy (2010) and his criticism of their heteronormative tenet (2012) – also questioned by Lauren Stephen (2010) in her queering of *March of the Penguins* – have paved the way for further research into filmmaking and animal ethics. Wildlife filmmaker Chris Palmer (2010, 2015) has been particularly vocal about the unethical and exploitative practices that are standard in the industry, and which affect both the production process and the representation of nonhuman animals. Fortunately, advances in ethology and a growing understanding of the threat of global warming, the Anthropocene and the Sixth Mass Extinction have gradually led to more ethical ways of representation and filmmaking (even by companies and channels that, on the other hand, also endorse other exploitative productions for the sake of ratings). It is these films and series, with their outspokenness about nonhuman animal sentience and their active call for conservation and environmentalism, which sometimes blur the boundaries with animal advocacy documentaries, as they also promote the preservation of ecosystems and biodiversity.

In addition, some conservationists' adamant position that human intervention in hazardous situations for wildlife entails artificiality and disruption is normalized as proper environmentalism. This is especially applicable in circumstances that qualify as part of the natural dynamics of life in the wild. In the Swiss film *Der Bär in mir* [*Bear-Like*] (Droux 2019), biologist David Bittner foreshadows the imminent death of a cub but does not contemplate possible assistance. In *My Octopus Teacher*, Craig Foster also restrains himself from intervening when the octopus he is bonding with is threatened by predators, for such are the natural cycles of the kelp forest. These productions, where animal advocacy is still prevalent, operate on conservationist convictions that pull the documentary toward the natural history/wildlife subgenre precisely because the principle of nonintervention is typical of such films. As such, it is common for wildlife documentaries to completely dispense with the appearance of humans onscreen.

Furthermore, in wildlife documentaries, human threat appears to be more scattered in terms of accountability. All humankind contributes to global warming and pollution, and so we are called upon, as a species, to change our habits for both our sakes and that of planetary biodiversity. In animal advocacy documentaries, dispersion of accountability is also common: most humans are consumers of animal products and by-products, and so it is our moral responsibility to cease such consumption. However, because in animal advocacy films we have also witnessed or have been persuasively informed about animal cruelty, our anagnorisis and feelings of guilt will be sharper in degree – we will become ashamed of our actions. In addition, this shared accountability is often coupled with an identifiable antagonist who might come in the form of a face, a name or a corporate logo. In *Blackfish*, for instance, SeaWorld is targeted and exposed as a business that is negligent toward orcas and trainers alike. As part of the rhetorizing of the conflict, the fact that executives or representatives of the oppressor decline to be interviewed or appear on film is seen as evidence of shame or lack of repentance. Lack of cooperation is seldom read on a neutral note.

Other types of films about nonhuman animals frame them as cultural assets. Tim Delmastro's *Cat Nation: A Film about Japan's Crazy Cat Culture* (2017) borrows from the ethnographic genre the trope of the shocked or curious Westerner "discovering" the value of a given symbol, totem or tradition in a remote culture. This knowledge is then retrieved and packaged for Western viewers. And because the quirks and antics of cats are so much a part of their representation in Western culture anyway, the film's charm lies in its levity and comic allowances. *True Appaloosa* (Woodman 2015) also repeats some conventions of the ethnographic genre, as it follows 69-year-old horse breeder Scott Engstrom into the heart of Kyrgyzstan in search of her "holy grail": pureblood Appaloosas that may "save" the breed from extinction and bring irrefutable evidence of her lifelong conviction that they originated in Central Asia. Yet for all the love professed for Appaloosas and their cultural irreplaceability, the obsession with purism (versus the "pollution" or "dirtiness" of crossbreeding) and the business interests at stake impede any proper conservationist discussion centered on animal ethics. The horses' interests are never truly considered – even arguably cruel sports such as *kok boru* are celebrated for their "amazing sportsmanship." Instead, the narrative focuses on Engstrom's emotional journey paralleling her travels, and Appaloosas are reinstated as assets only insofar as their blood is pure and they comply with breed standards. Breed fancy is rarely tolerated in animal ethics, as theorists and activists tend to associate aesthetic and temperamental expectations that people have of breeds with a desire for class status or group identity.

The consideration of nonhuman animals as cultural assets can, however, intersect with environmentalist and animal advocacy discourses. Films in this line seek a more holistic representation of nonhuman animals, one that assimilates the stories and images that make them part of the aesthetic (and often, foundational) identity of a culture. These cultural values and images are interlinked with beliefs about shared vulnerabilities and affective commonalities. Shaunak Sen's beautifully composed *All That Breathes*, a film about two brothers who rescue birds of prey in an environmentally hostile Delhi, poetically conveys the aesthetic and spiritual value of these birds in an urban landscape that is at once decayed and mesmerizing. They represent part of the Indian identity, at a historical and industrial time in which, in the words of one of the rescuers, "birds are plummeting from the sky. Delhi is a gaping wound. And we are a tiny Band-Aid on it." In these films, because of the belief in the continuum between nature and culture, conservation is inseparable from the reassertion of cultural identity.

One last type of subgenre that should be mentioned is the environmental documentary, often terminologically nuanced by the prefix "eco-" (the focus is on the systemic interconnections between species and the environment, structured into philosophical discussions) or the adjective "green" (more invested in practical, sustainable solutions and in educating the public about their choices). These films frequently overlap with animal advocacy documentaries: they are forms of media addressing local, grassroots and global issues, articulating conflicts that push ethics to the forefront. Environmentalism and animal rights share, after all, their desire to end anthropocentrism (Kemmerer and Kirjner 2015). Indeed, because of the conservationist message that is transversal to these subgenres, some critics integrate animal advocacy within the broader environmentalist ambit. Helen Hughes, for instance, parallels Louie Psihoyos's award-winning *The Cove* to Davis Gugghenheim's *An Inconvenient Truth* (2006) in terms of scope and narrative technique. Both films link the biography of an advocate (former trainer-turned-activist Ric O'Barry and former Vice President of the United States Al Gore, respectively) to the issue at stake. Their personal crises and dedication frame the representation of the conflict, as the revelations that have instigated their change make them relatable figures and legitimize their position as authorities on the subject (Hughes 2014, 119).

The frequent overlap and strong parallelisms between environmental and animal advocacy documentaries, both of which have proliferated in the twenty-first century, hinder any definitive, categorical distinction between the subgenres – a classification further problematized by the inherent plasticity of nonfiction film itself. Tracing the differences must be done on a case-by-case basis, and the focus must lie on the film's preferred framing of humans' moral obligations to nonhumans. In many documentaries targeting industrial farming and concentrated animal feeding operations (CAFOs), the welfarist perspective on animal exploitation is lukewarm. For example, Robert Kenner's *Food, Inc.* denounces the industry's corporate apparatus and its ruthless inconsideration of consumers and expendable workers, but the film reduces the consideration of nonhuman animal victims to a fraction of the argument. Welfarism (the idea that the suffering of nonhuman animals should be minimized, but their harvesting as food is not inherently wrong) prevails over animal rights perspectives deeming any form of exploitation morally unacceptable. In contrast, there are films that make it a point to represent nonhuman animals as sentient beings who suffer greatly throughout their short, horrific lives in these factory farms. And sometimes, the distinction between environmentalism and animal advocacy is more diluted. Where to draw the line (one could even question the need to draw one at all) in films such as *My Octopus Teacher*, *Sharkwater* (Stewart 2006) or *Sharkwater Extinction* (Stewart 2018) is difficult: the representation of the octopus and the sharks appearing in these films rests on a holistic understanding of the ecological need to preserve oceans and their biodiversity. The conservation of these species is a metonymy of a wider struggle of planetary proportions. And, particularly in *My Octopus Teacher*, the portrayal of the octopus not as a specimen but as an intelligent, sentient individual, foregrounds moments of interspecies intimacy that spark the types of affects sought by animal advocacy documentaries.

History and Archival Practice

The films discussed in this book are indebted to the renewed interest in documentaries that emerged at the beginning of the twenty-first century (Rich 2006; Smaill 2016, 2) and that was characterized by successful theatrical release, box-office profitability and the strategic commercialism of film distributors. Historiographic reconstructions of this renaissance often point to films from the early 2000s as samples of the landmark productions that reignited public enthusiasm for documentaries as sources of knowledge and entertainment. Ranking high in this list are *March of the Penguins* and *An Inconvenient Truth*, along with other productions such as Hubert Sauper's *Darwin's Nightmare* (2004), Michael Moore's *Bowling for Columbine* (2002) and *Fahrenheit 9/11* (2004), and Werner Herzog's *Grizzly Man* (2005) and *Encounters at the End of the World* (2007). Although, as established at the beginning of this introduction, many animal advocacy documentaries have not been circulated by mainstream outlets, the subgenre has surely benefited from the popularity and public taste for nonfiction film in these past decades. Claire Parkinson and Lara Herring specifically attribute twenty-first-century interest in animal advocacy documentary film to the appearance of Shaun Monson's *Earthlings* in 2005. "Prior to this," they write, "animal rights films were often low- or 'no-' budget productions, usually involving undercover filming, with short run times and produced predominantly by animal advocacy groups with little in the way of formal distribution routes open to them" (2024, 1). *Earthlings* became a game changer, proving that "a feature-length animal rights film could have an extensive reach" (2024, 1), and marking the beginning of a wave of similarly oriented productions that embraced "new opportunities for financing and distribution [that had] become more readily available" (2024, 2).

It could be argued that the animal advocacy documentaries of the early 2000s helped the animal rights movement to recover from a decade of fatigue and confusion that had brought activists to a standstill in the public eye. Harold Herzog took note of the crisis in a short article titled "Has Public Interest in Animal Rights Peaked?," published in *American Psychologist* in 1995. Following the premise that public concern for social issues can be quantitatively measured by the amount of media coverage they get, Herzog argued that "the visibility of the animal rights movement has leveled off and may be declining" (1995, 946), and provided data suggesting that the consumption of fur and meat in the United States had been rising for the last two years. The wane of public interest, however, did not necessarily mean animal rights activism was disappearing, but that other types of strategies "away from the barricades and toward courtrooms and statehouses" (Herzog 1995, 947) had become the priority, especially after reforms such as the 1985 Animal Welfare Act. A follow-up comment published in *Science* chimed in on this hypothesis by quoting Frankie Trull, of the Foundation for Biomedical Research. According to Trull, animal rights activism had become more "insidious" (quoted in Holden 1929, 1995), infiltrating public and education systems in more subtle ways.

Historians of animal activism in the United States became aware of this crisis of the mid-1990s very early on. Lawrence Finsen and Susan Finsen's well-known *The Animal Rights Movement in America*, published in 1994, already noted that the movement stood at a "critical crossroads" (1994, 257) that reflected internal disagreements about ethical direction (reform or abolitionism), effective activism and the prioritization of specific issues. More recently, Emily Patterson-Kane, Michael Allen and Jennifer Eadie have described the movement at the turn of the twenty-first century as the "scattered debris" of "an explosion of theoretical and practical activity concerning animal rights" (2022, 133). They refer to the so-called "theory fatigue" (2022, 136) that evinced activists' impatience with the academic discordance between abolitionism and welfarism and that ultimately gave rise to the promotion of a culture of responsible consumerism in which philosophical conviction became less central. The resurgence of the documentary genre, therefore, came at a time when uncertainty about the identity and strategies of the movement made room for a re-exploration of the possibilities afforded by the fast-changing media and its communicative potential. If theory and ambivalence toward the welfare-or-rights spectrum had diminished the visibility of the movement in the public sphere, filmmakers certainly helped to reignite discussions with their circulation of images and filmic ekphrases representing nonhuman animals as sentient, self-aware beings.

In addition, the role of these films in the audiovisual reinvigoration of the movement serves a historiographic purpose, as they chronicle, write and rewrite the history of nonhuman animals within human cultures and the animal-industrial complex. They constitute archival renditions that cleanse dominant historical narratives, freeing nonhuman animals from anthropocentric bias and framing them as victims. However, historiographic methods that rehabilitate nonhuman animals and carve a space of memory for them are not without their epistemological problems. Erica Fudge argues that the only way to truly articulate the history of nonhuman animals is through antihumanism: "By refusing humanism, and, implicitly, anthropocentrism, we place ourselves next to the animals, rather than as users of the animals . . . The history of animals can only work at the expense of the human" (2002, 15). And yet the documentary genre cannot avoid audiovisual representation or the creation of a narrative; the anthropocentric nature of its formal semiotics curtails the possibility of historicizing nonhuman animals beyond the archiving of humanist thought. Their narratives of truth are composed by intertextualities built by and for humans. As Fudge acknowledges prior to her optimism toward antihumanism, the history of nonhuman animals "is in reality the history of human attitudes toward animals," itself just "a part of the history of ideas: nothing really new at all. . . . [W]e are never looking at the animals, only ever at the representation of the animals by humans" (2002, 6).

Aside from the question of whether antihumanism is realistically achievable (and what this would look like in absolute terms),[4] there is also the matter as to whether it is desirable. As humans, are we not more sensitized to human paradigms (of morality, for instance) than we are to nonhuman ways of experiencing and making sense of the world? Does not anagnorisis and the recognition of our morally reprehensible actions create a catharsis likely to result from effective humanist forms of storytelling? By humanist I do not mean that humans are to be the primordial moral concern of the narrative, which would again objectify nonhumans. I mean that the choices affecting narrativization, dramatization, montage, editing etc. are intrinsically humanistic, as is the very art of persuasion. These aesthetic possibilities can be put to the service of representing nonhuman subjectivity in effective ways, insofar as effectiveness is measured by the ability to elicit transformative change.

In relation to this discussion, it is important to emphasize the role of the filmmakers behind animal advocacy documentaries as not just storytellers: they are authorial and authoritative voices that forge and arrange the imagery and narratives shaping our collective memory as cultures connected to nonhuman otherness. Belinda Smaill (2016) has pioneered outstanding work on the archival character inherent in how the documentary moving image registers nonhuman animal bodies. Specifically, I would argue that animal advocacy films can be regarded as community archives, as they "characterise a non-traditional archival collection specifically tied to a particular group, often one that may be undocumented or under-documented by traditional archival institutions" (Bastian and Flinn 2020, xx). It is essential to stress the role of filmmakers (alongside photographers, journalists and even nonprofessional contributors to digital social and news media) as archivists who create the audiovisual material necessary for the articulation and consolidation of a collective memory that acknowledges and grants visibility to nonhuman others, who have been historically marginalized and ignored by memory studies. By documenting, preserving and sharing the corpus of images and sounds that record the instrumentalization of nonhuman animals in a way that represents them as the sentient subjects they are, archivists engage in mnemonic methodologies that disrupt anthropocentrism.

Some institutions have launched much-needed archives of their own, such as the Ryder Papers, held at the British Library, or the Tom Regan Animal Rights Archive at the University of North Carolina Wilmington. But these collections are mostly premised on documenting the history of the animal rights movement. Catherine Oliver writes that the Ryder Papers were "primarily preserved to document the emergence of the politically and socially important animal activism movement in Britain: a movement of humans" (2022, 21). This reflects Fudge's claim that the history of nonhuman animals has essentially been framed as a history of human attitudes toward them. A different type of archival practice – one that more closely resembles the endeavor of animal advocacy documentaries – is that of We Animals, a photojournalism agency founded by Jo-Anne McArthur that documents and collects (audio)visual evidence of animal exploitation worldwide, as well as more positive images of rescues and sanctuaries. We Animals employs strategic networking to ensure that the images and videos reach the public and can be used by animal rights organizations. As of today, more than 40,000 stock photos and videos are available on their website.

This is not to say that animal advocacy documentaries discard thematic engagements with the history of the movement or the journey of advocates and activists. These are, in fact, central concerns in many of the films. But when it comes to their original contribution as an archive, their merit arguably lies in provisioning a corpus and an intertext that reflect the aesthetic, audiovisual ranges that have historically been deployed to convey the "truth" of nonhuman animal experience in the animal-industrial complex. Aside from that, the subgenre also has a lot to offer to the historicization of the ideas and attitudes toward nonhuman animals. These historiographic revisions are often supplemented by utopic and dystopic visions of what lies ahead, as "political engagement of memory work and its transformative potential involves not only publicly dealing with past atrocities but also imagining new futures" (Gutman and Wüstenberg 2023, 9).

Lastly, it is important to note that the documentary-text-as-intertext does not just nourish the archival impulse to re-historicize nonhuman animals. It also points to the awareness of a tradition in the making, as films borrow thematic and formal material from other films, which is then rearranged and infused with new (aesthetic) meaningfulness. This open, organic expansion of the subgenre operates according to intertextual rules that keep it alive, situating new works that imitate and respond to previous ones, and leading not just to a growth in numbers but to the evolution of an aesthetic.

Scope and Structure of This Book

The following chapters are conceived as analyses of documentaries whose representativeness within the animal advocacy subgenre invites a closer inspection of the formal and thematic conventions that characterize them, and of how filmmakers infuse them with their own creative vision. While the chapters implicitly serve to legitimize the subgenre, it is not the object of this book to gravitate toward mere descriptive expositions of its features. Instead, the analyses offer critical approaches to problematize how (audio)visual indexes of the moving image, along with other factors pertaining to form and content, are deployed with the aim of prompting affective responses on the part of viewers. As such, the book fundamentally employs approaches from the field of the humanities, focusing on an exegesis informed by the arts of representation and persuasion, and limiting discussions of the subgenre's place in the market of documentary commons. The book's central interest in the shaping of documentary narratives and their underlying authorial pulse demands a practical application of theoretical frameworks that intersect with animal advocacy discourses. These frameworks engage with issues involving identity politics, the ethics of (in)visibility, the underlying forces of genre conventions, the celebrity apparatus and historiographic revisionism.

Regarding the corpus, most of the films analyzed in the forthcoming chapters are British or American (by which I mean United States) productions. This is not just a reflection of the canonical dominance of English-language films (Australian productions included) but also a sign of the central role that Great Britain and the United States have historically played in modeling animal ethics discourses and narratives with universal claims. Because of the international impact of these films – thanks to theatrical and/or online distribution, critical acclaim at festivals, academic interest and strategic circulation by activists – it is fair to say that they largely embody the precepts of Western aesthetics and sensibilities. However, Spanish, French and German films will also occasionally be drawn into the discussion to further critical commentary and refine the Western take on the subgenre.

The documentaries under discussion are mostly feature-length, though shorter formats are also assessed in some sections for argumentative purposes. The criterion for feature length varies depending on the institution setting the standards – for instance, the Academy of Motion Picture Arts and Sciences establishes a minimum of 40 minutes, and the British Film Institute requires more than 60 minutes. I am less interested in fixing a minimum running time than I am in considering a general threshold for these films, which more closely aligns with the BFI's guidelines in this regard. Even if they only represent a portion of animal advocacy media, longer narratives provide more time for expositional development and rhetorical recursiveness and complexity, often increasing the chances of producing a meaningful affective response in viewers. A longer duration can more elaborately create a sense of intimacy with onscreen subjects and provide more sustained immersive cinematographic techniques. Feature-length films are also more extensively circulated in movie theaters, festivals and streaming platforms, and those that do not enjoy the exposure of mainstream cinematic outlets often find a reliable means of distribution on YouTube and other similar services, which garners them a sizable audience. As mentioned earlier, this study does not resort to commercial success (or accolades) as a criterion for the corpus, for it stands firm in the belief that this subgenre's canon owes its identity to both cinematic productions and more independent, low-budget films, all of which conform to the intertextual praxis.

It must also be clarified that this book employs the term "animal advocacy" documentary as opposed to other commonly used modifiers (animal activist, animal rights, animal liberation) to encompass a full spectrum of ethical stances. Although, as argued earlier, the subgenre is strongly rooted in activist filmmaking, and although it is my contention that all the films seek to inspire some form of transformative change, I am wary of labels that might implicitly make assumptions about authorial intent or a film's rationale. Filmmakers such as Victor Kossakovsky and Andrea Arnold have openly resisted categorizations that could reduce their work to vegan "propaganda," manifesting a desire for their films to be considered as something else (or at least as something beyond). To my knowledge, there is also no evidence of a commitment to animal rights on the part of director James Marsh, despite the criticism of animal experimentation in *Project Nim*. In addition, films such as *The Game Changers* (Psihoyos 2018) and *The Invisible Vegan* (Perry 2019) mostly base their arguments on plant-based nutrition in ways that do not prioritize animal rights or liberation. Because of this, "advocacy" appears to be the more appropriate term for referring broadly to films that take an ethical stance directly or indirectly favoring nonhuman animals, that on some level critique the capitalist logics of the animal-industrial complex and that address the range of positions and practical strategies across the spectrum of animal ethics, from more liberationist to more welfare-oriented approaches. When the term "animal activist documentary" is employed, it is because the films under discussion at that point do unequivocally invoke an activist stance.

Chapter 1 examines the tactical use of moral shock in animal advocacy film – one of the most salient strategies employed by advocates and activists. It addresses the ethical and aesthetic implications of graphic representations of violence inflicted on nonhuman animal bodies, and how this signifies the dichotomization between what is visible and what is concealed by the animal-industrial complex. The quantitative and qualitative treatment of violence in Shaun Monson's *Earthlings* (2005) and Chris Delforce's *Lucent* (2014) and *Dominion* (2018) seek an embodied response on the part of the viewer, and they integrate the image of the suffering nonhuman animal within the politics of historicization.

Chapter 2 explores the techniques used by filmmakers to trace and represent the lives of singled-out nonhuman individuals. I propose the term "biograph-able subject" to reflect on the elements that cohesively shape a life narrative and a conflict marked by the themes of trauma and maternal deprivation. The anti-captivity narratives in James Marsh's *Project Nim* (2011) and Gabriela Cowperthwaite's *Blackfish* (2013) evince the tensions between the anthropocentric forces of the biographical genre, which seek authenticity through the string-of-interviews technique, and an ethical representation of the individual meant to function synecdochically as a symbol of their species. To deal with this tension and the problematic issues it raises regarding the "(un)knowability" of the individual protagonist, I turn to Martha Nussbaum's (2022) proclamation of the feeling of wonder as an ethically sound response – one that moving-image representations of nonhuman animals can inspire.

Chapter 3 identifies and analyzes common paradigms employed by the subgenre to justify and arrange the moral actions and beliefs embodied by activists and advocates. The moral guidance offered by these figures rests on a dichotomization of forms of agency that is patterned in stories of trials and tribulations. The chapter begins with an examination of the conversion narrative in Jenny Stein's *Peaceable Kingdom: The Journey Home* (2009) and Marc Pierschel's *Butenland* (2019). It then discusses the deconstruction of antiterrorist rhetoric in post-9/11 documentaries such as Shannon Keith's *Behind the Mask* (2006), Curt Johnson's *Your Mommy Kills Animals* (2007) and Casey Suchan and Denis Henry Hennelly's *The Animal People* (2019). The final section addresses the framing of discourses of advocacy and activism in relation to racial and cultural identity, and the problematic tensions this can create in documentary representation, with particular attention to Jasmine Perry's *The Invisible Vegan* (2019) and Louie Psihoyos's *The Cove* (2009).

In view of the extensive scholarship on the connections between speciesism and sexism, Chapter 4 explores how gender studies can inform and guide the interpretation of animal advocacy documentaries. These documentaries emerge as rich sites for critiquing the politics of normative masculinity, but they can also engage problematically with dominant expectations of hypermasculinity – as depicted in Louie Psihoyos's *The Game Changers* (2018) – and reaffirm strategic, recurrent tropes employed by men in their filmmaking. Contesting these conventions, films such as Liz Marshall's *The Ghosts in Our Machine* (2013), Andrea Arnold's *Cow* (2021) and Victor Kossakovsky's *Gunda* (2020) invite readings of creative and cinematographic choices from the perspective of what Lisa French (2022) calls the "female gaze" and the "feminine aesthetic."

Chapter 5 addresses the subgenre's instrumentalization of the celebrity apparatus and the types of capital that celebrities can leverage when participating in these documentaries. These collaborations involve risks for both advocates and celebrities, evincing the delicate balance of credibility, authority and authenticity in moral standing. As manufactured discourses, celebrities lend their talents, faces and/or bodies to bring visibility to animal rights issues and (ethical) veganism. Their endorsement requires negotiating forms of "consensualism" that safeguard their public favorability. I examine examples of celebrity collaborations in these films, with a particular interest in Shaun Monson's *Unity* (2015), before turning to a case study of the activist trajectory of one of his most frequent collaborators, Joaquin Phoenix.

Finally, Chapter 6 engages with post-postmodernism as an aesthetic approach that overcomes the perceived stagnancy of postmodernism and embraces ethical self-reflexivity. In the context of the Anthropocene, of which the animal-industrial complex is a prime cause, filmmakers explore the formal and thematic mechanisms at the service of documentary to devise narratives of hope for the future of humanity and biodiversity, or to propose alternative forms of species coexistence that revisit the spectrality of nonhuman otherness in modernity. Through the analysis of three films – Simon Amstell's *Carnage: Swallowing the Past* (2017), Louie Psihoyos's *Racing Extinction* (2015) and Emilio Fonseca's *Salvaxe, salvaxe* (2024) – I contend that the mockumentary genre and the trope of (digital) technology, as both theme and form, are evidence of post-postmodernist ethical self-reflexivity.

Notes

1 The research for this book was conducted as part of two projects – "The Ethics and Aesthetics of Animal Advocacy Documentary Film in Twenty-first-century Western Cultures" (CIAICO/2023/046) and "Representations of Masculinities in Animal Advocacy Documentaries in English (2000–2021)" (CIGE/2021/100) – funded by the Conselleria d'Educació, Cultura i Universitats (Generalitat Valenciana).

2 A note on the text: the term "nonhuman animal(s)" will be used throughout this book to comply with one of the semantic preferences established in the field of Critical Animal Studies. However, to avoid excessive wordiness in the discussion of concepts, nominal compounds in the line of "animal rights," "animal advocacy," "animal activism," "animal exploitation," "animal-industrial complex" etc. will remain untouched, even if such terms specifically refer to nonhuman animals.

3 Translations into English of the newscast voice-overs have been made by the author.

4 Phenomenological interest regarding the possibility of experiencing or fully apprehending nonhuman awareness of the world was famously explored by Thomas Nagel in his essay "What Is It Like to Be a Bat?" (1974). Using bats' idiosyncratic form of sensory experience (echolocation) as an example, Nagel concluded that it is ultimately impossible for humans to know bats' subjectivity (i.e., we cannot know what it is like for a bat to be a bat). The closest that we can come to bat consciousness is by imagining what it would be like for a human to be a bat – an experiential form that requires human-centered representation.

References

Adams, Carol J. 1990. *The Sexual Politics of Meat. A Feminist-Vegetarian Critical Theory*. Continuum.

Aguayo, Angela J. 2019. *Documentary Resistance. Social Change and Participatory Media*. Oxford University Press.

Almiron, Núria. 2016. "The Political Economy Behind the Oppression of Other Animals. Interest and Influence." In *Critical Animal and Media Studies. Communication for Nonhuman Animal Advocacy*, edited by Núria Almiron, Matthew Cole, and Carrie P. Freeman, 26–41. Routledge.

Almiron, Núria, Matthew Cole, and Carrie P. Freeman, eds. 2016. *Critical Animal and Media Studies. Communication for Nonhuman Animal Advocacy*. Routledge.

Amstell, Simon, dir. 2017. *Carnage: Swallowing the Past*. BBC.

Arnold, Andrea, dir. 2021. *Cow*. IFC Films, BBC Film.

Aufderheide, Patricia. 2007. *Documentary Film. A Very Short Introduction*. Oxford University Press.

Baker, Steve. (1993) 2001. *Picturing the Beast. Animals, Identity, and Representation*. University of Illinois Press.

Bastian, Jeanette A., and Andrew Flinn. 2020. "Introduction." In *Community Archives, Community Spaces: Heritage, Memory and Identity*, edited by Jeanette A. Bastian and Andrew Flinn, xix–xxiv. Facet Publishing.

Berger, John. (1980) 2009. *Why Look at Animals*? Penguin.

Bousé, Derek. 2000. *Wildlife Films*. University of Pennsylvania Press.

Burgos, Javier, dir. 1986. *Hidden Crimes*. SUPRESS.

Burt, Jonathan. 2005. "Review of *A Day in the Life of a Massachusetts Slaughterhouse* (PETA Videos/1998)." *Society & Animals* 13 (4): 347–54. https://doi.org/10.1163/156853005774653663.

Chapman, Jane L. 2009. *Issues in Contemporary Documentary*. Polity Press.

Chris, Cynthia. 2006. *Watching Wildlife*. University of Minnesota Press.

Corner, John. 1996. *The Art of Record. A Critical Introduction to Documentary*. Manchester University Press.

Corner, John. 2002. "Performing the Real: Documentary Diversions." *Television & New Media* 3 (3): 255–69. https://doi.org/10.1177/152747640200300302.

Cowperthwaite, Gabriela, dir. 2013. *Blackfish*. Magnolia Pictures.

Delforce, Chris, dir. 2014. *Lucent*. Aussie Farms.

Delforce, Chris, dir. 2018. *Dominion*. Aussie Farms.

Delmastro, Tim, dir. 2017. *Cat Nation: A Film About Japan's Crazy Cat Culture*. 100th Monkey Films, Moving Screen, Journeyman Pictures.

Droux, Roman, dir. 2019. *Der Bär in mir*. Memox GmbH, Journeyman Pictures.
Dunayer, Joan. 1995. "Sexist Words, Speciesist Roots." In *Animals and Women: Feminist Theoretical Explorations*, edited by Carol Adams and Josephine Donovan, 11–31. Duke University Press.
Duvall, John A. 2017. *The Environmental Documentary: Cinema Activism in the 21st Century*. Bloomsbury.
Ehrlich, Pippa, and James Reed, dirs. 2020. *My Octopus Teacher*. Off the Fence, The Sea Project, Netflix.
Etzioni, Amitai. 1970. *Demonstration Democracy*. Gordon and Breach Science Publishers.
Farm Sanctuary. 1991. *The Down Side of Livestock Marketing*.
Finsen, Lawrence, and Susan Finsen. 1994. *The Animal Rights Movement in America. From Compassion to Respect*. Twayne.
Flaherty, Robert, dir. 1922. *Nanook of the North*. Revillon Frères, Pathé Exchange.
Fonseca, Emilio, dir. 2024. *Salvaxe, salvaxe*. Walkie Talkie Films.
Freeman, Carrie Packwood, and Scott Tulloch. 2013. "Was Blind But Now I See: Animal Liberation Documentaries' Deconstruction of Barriers to Witnessing Injustice." In *Screening Nature. Cinema Beyond the Human*, edited by Anat Pick and Guinevere Narraway, 110–26. Berghahn.
French, Lisa. 2022. *The Female Gaze in Documentary Film. An International Perspective*. Palgrave Macmillan.
Fudge, Erica. 2002. "A Left-Handed Blow: Writing the History of Animals." In *Representing Animals*, edited by Nigel Rothfels, 3–18. Indiana University Press.
Godmilow, Jill. 1997. "How Real Is the Reality in Documentary Film? Interview by Ann-Louise Shapiro." *History and Theory* 36 (4): 80–101. https://doi.org/10.1111/0018-2656.00032.
Goode, Eric, and Rebecca Chaiklin, dirs. 2020. *Tiger King: Murder, Mayhem and Madness*. Season 1. Netflix.
Gruen, Lori, comp. 1986. *Britches*. PETA.
Guggenheim, Davis, dir. 2006. *An Inconvenient Truth*. Participant Productions, Paramount Classics, Lawrence Bender/Laurie David.
Gutman, Yifat, and Jenny Wüstenberg. 2023. "Introduction: The Activist Turn in Memory Studies." In *The Routledge Handbook of Memory Activism*, edited by Yifat Gutman and Jenny Wüstenberg, 5–15. Routledge.
Herzog, Harold A. 1995. "Has Public Interest in Animal Rights Peaked?" *American Psychologist* 50 (11): 945–47. https://doi.org/10.1037/0003-066X.50.11.945.
Herzog, Werner, dir. 2005. *Grizzly Man*. Lions Gate Films, Discovery Docs, Real Big Productions.
Herzog, Werner, dir. 2007. *Encounters at the End of the World*. Discovery Films, Creative Differences Productions.
Holden, Constance, ed. 1995. "Animal Rights Fading?" *Science* 270 (5244): 1927, 1929. https://doi.org/10.1126/science.270.5244.1927.
Horkheimer, Max, and Theodor W. Adorno. (1947) 2002. *Dialectic of Enlightenment: Philosophical Fragments*. Edited by Gunzelin Schmid Noerr. Translated by Edmund Jephcott. Stanford University Press.
Hughes, Helen. 2014. *Green Documentary. Environmental Documentary in the 21st Century*. Intellect.
Hunnicut, Gwen, Richard Twine, and Kenneth Mentor. 2025. "Towards Multispecies Justice: Unveiling Violence and Exploitation in the Animal-Industrial Complex." In *Violence and Harm in the Animal Industrial Complex*, edited by Gwen Hunnicut, Richard Twine, and Kenneth Mentor, 1–17. Routledge.
Jacquet, Luc, dir. 2005. *La Marche de l'empereur*. Bonne Pioche, Wild Bunch, Warner Independent Pictures, National Geographic.
Johnson, Curt, dir. 2007. *Your Mommy Kills Animals*. Indie Genius Productions.
Keith, Shannon, dir. 2006. *Behind the Mask*. Uncaged Films, ARME.
Kemmerer, Lisa, and Daniel Kirjner. 2015. "Conflict and Accord. A Critical Review of Theory and Methods for Earth and Animal Advocacy." In *Animals and the Environment. Advocacy, Activism, and the Quest for Common Ground*, edited by Lisa Kemmerer, 15–38. Routledge.

Kenner, Robert, dir. 2008. *Food, Inc.* Magnolia Pictures, Participant Media, River Road Entertainment.
Kim, Jihoon. 2022. *Documentary's Expanded Fields: New Media and the Twenty-First-Century Documentary*. Oxford University Press.
Kossakovsky, Victor, dir. 2020. *Gunda*. Sant & Usant, Louverture Films.
Lévi-Strauss, Claude. 1964. *Totemism*. Translated by Rodney Needham. Merlin Press.
Lippit, Akira Mizuta. 2000. *Electric Animal. Towards a Rhetoric of Wildlife*. University of Minnesota Press.
Malamud, Randy. 2012. *An Introduction to Animals and Visual Culture*. Palgrave Macmillan.
Marsh, James, dir. 2011. *Project Nim*. BBC Films, UK Film Council, HBO Documentary Films, Red Box Films, Passion Pictures, Roadside Attractions.
Marshall, Liz, dir. 2013. *The Ghosts in Our Machine*. Ghosts Media, LizMars Productions.
Merskin, Debra. 2016. "Media Theories and the Crossroads of Critical Animal and Media Studies." In *Critical Animal and Media Studies. Communication for Nonhuman Animal Advocacy*, edited by Núria Almiron, Matthew Cole, and Carrie P. Freeman, 11–25. Routledge.
Mills, Brett. 2010. "Television Wildlife Documentaries and Animals' Right to Privacy." *Journal of Media & Cultural Studies (Continuum)* 24 (2): 193–202. https://doi.org/10.1080/10304310903362726.
Mills, Brett. 2012. "The Animals Went in Two by Two: Heteronormativity in Television Wildlife Documentaries." *European Journal of Cultural Studies* 16 (1): 100–14. https://doi.org/10.1177/1367549412457477.
Mills, Brett. 2017. *Animals on Television. The Cultural Making of the Non-Human*. Palgrave Macmillan.
Molloy, Claire. *Popular Media and Animals*. Palgrave Macmillan, 2011.
Monson, Shaun, dir. 2005. *Earthlings*. Nation Earth.
Monson, Shaun, dir. 2015. *Unity*. Nation Earth.
Moore, Michael, dir. 2002. *Bowling for Columbine*. Dog Eat Dog Films, Salter Street Films, United Artists, Alliance Atlantis.
Moore, Michael, dir. 2004. *Fahrenheit 9/11*. Dog Eat Dog Films, Fellowship Adventure Group, IFC Films, Lionsgate Films.
Nagel, Thomas. 1974. "What Is It Like to be a Bat?" *Philosophical Review* 83 (4): 435–50. https://doi.org/10.2307/2183914.
Newkirk, Ingrid, and Alex Pacheco, prods. 1984. *Unnecessary Fuss*. PETA.
Nibert, David. 2002. *Animal Rights. Human Rights. Entanglements of Oppression and Liberation*. Rowman and Littlefield.
Nibert, David. 2013. *Animal Oppression and Human Violence. Domesecration, Capitalism, and Global Conflict*. Columbia University Press.
Nichols, Bill. 2001. *Introduction to Documentary*. Indiana University Press.
Noske, Barbara. 1989. *Humans and Other Animals. Beyond the Boundaries of Anthropology*. Pluto Press.
Nussbaum, Martha C. 2022. *Justice for Animals. Our Collective Responsibility*. Simon & Schuster.
Nuwer, Rachel. 2020. "Why *Tiger King* Is Not *Blackfish* for Big Cats." *The New York Times*, April 9. https://www.nytimes.com/2020/04/09/science/tiger-king-joe-exotic-conservation.html.
Oliver, Catherine. 2022. *Veganism, Archives and Animals. Geographies of a Multispecies World*. Routledge.
Palmer, Chris. 2010. *Shooting in the Wild*. Sierra Club.
Palmer, Chris. 2015. *Confessions of a Wildlife Filmmaker*. Bluefield Publishing.
Parkinson, Claire, and Lara Herring. 2024. "Introduction." In *Animal Activism On and Off Screen*, edited by Claire Parkinson and Lara Herring, 1–12. Sydney University Press.
Patterson-Kane, Emily, Michael P. Allen, and Jennifer Eadie. 2022. *Rethinking the Animal Rights Movement*. Routledge.
Perry, Jasmine C., dir. 2019. *The Invisible Vegan*. Kenny & Jasmine Leyva Films.

Phelps, Norm. 2007. *The Longest Struggle. Animal Advocacy from Pythagoras to PETA*. Lantern Books.
Pick, Anat. 2018. "Vegan Cinema." In *Thinking Veganism in Literature and Culture. Towards a Vegan Theory*, edited by Emelia Quinn and Benjamin Westwood, 125–46. Palgrave Macmillan.
Pierschel, Marc, dir. 2019. *Butenland*. Blackrabbit Images.
Psihoyos, Louie, dir. 2009. *The Cove*. Ocean Preservation Society, Diamond Docs, Skyfish Films.
Psihoyos, Louie, dir. 2015. *Racing Extinction*. Okeanos, Discovery, Oceanic Preservation Society.
Psihoyos, Louie, dir. 2018. *The Game Changers*. ReFuel Productions, Oceanic Preservation Society, Diamond Docs.
Regan, Tom, dir. 1986. *We Are All Noah*. Culture & Animals Foundation.
Regan, Tom, dir. 1988. *Voices I Have Heard*. Culture & Animals Foundation.
Rich, B. Ruby. 2006. "Documentary Disciplines: An Introduction." *Cinema Journal* 46 (1): 108–15. https://doi.org/10.1353/cj.2007.0004.
Rothfels, Nigel, ed. 2002. *Representing Animals*. Indiana University Press.
Saget, Bob, dir. 2006. *Farce of the Penguins*. ThinkFilm, Permut Presentations, Two Angels Productions.
Sauper, Hubert, dir. 2004. *Darwin's Nightmare*. Mille et Une Productions, Coop99 Filmproduktion, Saga Film.
Schonfeld, Victor, and Myriam Alaux, dirs. 1981. *The Animals Film*. Slick Pics International.
Sen, Shaunak, dir. 2022. *All That Breathes*. HBO, Rise Films, Submarine Deluxe, Sideshow, HHMI Tangled Bank Studios.
Shukin, Nicole. 2009. *Animal Capital. Rendering Life in Biopolitical Times*. University of Minnesota Press.
Smaill, Belinda. 2016. *Regarding Life. Animals and the Documentary Moving Image*. State University of New York Press.
Stein, Jenny, dir. 2009. *Peaceable Kingdom: The Journey Home*. Tribe of Heart.
Stephen, Lauren C. 2010. "'At Last the Family Is Together': Reproductive Futurism in *March of the Penguins*." *Social Identities* 16 (1): 103–18. https://doi.org/10.1080/13504630903465944.
Stewart, Rob, dir. 2006. *Sharkwater*. Sharkwater Productions, Diatribe Pictures.
Stewart, Rob, dir. 2018. *Sharkwater Extinction*. Sharkwater Productions, Diatribe Pictures.
Suchan, Casey, and Denis Henry Hennelly, dirs. 2019. *The Animal People*. Finngate, Seafox, Birdstreet Productions, Virgil Films.
Telecinco. 2022. *Informativos Telecinco*, April 10. https://www.telecinco.es/noticias/a-la-carta/informativo-fin-semana-mediodia-domingo-10-04-2022_18_3311971183.html.
Trumbore, Dave. 2020. "*Tiger King* is Trash TV that Chooses Memes over Meaningful Action." *Collider*, April 2. https://collider.com/tiger-king-is-trash-tv.
Twine, Richard. 2012. "Revealing the 'Animal-Industrial' Complex': A Concept and Method for Critical Animal Studies?" *Journal for Critical Animal Studies* 10 (1): 12–39.
Vallejo, Aida. 2020. "Rethinking the Canon: The Role of Film Festivals in Shaping Film History." *Studies in European Cinema* 17 (2): 155–69. https://doi.org/10.1080/17411548.2020.1765631.
Whedbee, Karen E. 2019. "Reverend Billy Goes to Main Street: Free Speech, Trespassing, and Activist Documentary Film." *Journal of Film and Video* 71 (2): 30–46. https://doi.org/10.5406/JFILMVIDEO.71.2.0030.
Woodman, Conor, dir. 2015. *True Appaloosa*. Hustler's Trumpet.

1 Moral Shock and the Representation of Nonhuman Animal Bodies as Sites of Violence and Authenticity

This chapter addresses animal advocacy documentaries' representations of nonhuman animal bodies as sites where violence is systematically and repeatedly exerted. Graphic and excessive representation of violence is a common strategy employed by activists in (audio)visual media striving to produce moral shock in spectators. This, however, leads to a series of ethical and aesthetic considerations that point to a variety of possible cognitive and affective responses on the part of viewers. Through a review of the most salient scholarly literature on moral shock and the (in)visibility of nonhuman animal suffering, the analysis foregrounds how the compiling, selection and arrangement of images create carefully vertebrated films forced into an art of conviction that must address the meaningfulness of the representation of nonhuman animal bodies as authenticators of truths. After a brief critical overview of how such politics of sight are executed in a sample of earlier and/or shorter films, I turn to an in-depth discussion of the aesthetics employed by three canonical feature documentaries: Shaun Monson's *Earthlings* (2005), an American film, and Chris Delforce's *Lucent* (2014) and *Dominion* (2018), both Australian films. Contrary to the use of traditional plot structures around conflict (rising action, climax, denouement) in other documentaries, these three films are articulated by a compilatory syntax of persistent images of abused nonhuman animals and voice-over narration. Theirs is not a journey toward a delayed climax that pushes the viewer to a moment of anagnorisis, nor do they attempt to entertain. What they instead surrender, as archives of sorts, is a visual anthology of the horrific treatment of nonhuman animals in a (capitalist) animal-industrial complex specifically designed to harvest their bodies. Every stage in this manufacturing offers a grisly series of haunting images of suffering, pain and terror.

Building Shock

Despite the perceived dissatisfaction with the accomplishments to date of the animal rights movement (Arcari 2022), and the fact that moral shock tactics are frequently questioned by animal advocates and campaigners for their actual effectiveness (Mika 2006; Scudder and Mils 2009; Wrenn 2013; Fittock 2018; Fernández 2020), this type of strategy has become a common device in animal advocacy documentary filmmaking. The belief in the potential of shock tactics to bring about

DOI: 10.4324/9781032661445-2

transformative change and some level of reform influences filmmakers' arrangement of visual evidence, leading to stylistics that, in the name of effectiveness, combine techniques that affectively keep viewers on edge.

Bill Nichols's (2001) classification of documentary modes (expository, observational, participatory, reflexive, performative and poetic) provides a framework to approach films at a macro and/or micro level (Natusch and Hawkins 2014). Single films may draw on disparate cinematic techniques to maximize persuasion and absorb in their form of storytelling characteristics of different modes, leveling out the degree to which the camera and extradiegetic resources negotiate their informative and affective depiction of the "truth" behind nonhuman animal exploitation and the subjective experience of its victims. *Earthlings* and *Dominion* comply with the basic elements of the expository mode, although the observational and performative modes are also noticeably deployed. The expository mode "emphasizes verbal commentary and argumentative logic"; the observational mode "emphasizes direct engagement with the everyday life of subjects as observed by an unobtrusive camera"; and the performative mode "emphasizes the subjective or expressive aspect of the filmmaker's own engagement with the subject and an audience's responsiveness to this engagement" (Nichols 2001, 33–34).[1] *Lucent* arguably resorts more to the observational mode but also employs characteristics of the expository and performative ones. The expository mode brings in an extradiegetic narrator (usually characterized by a grave, heavy tone), and music augments the emotional and subjective engagement with the images, thus accentuating the performative mode. Though sentimental or ethereal lyricism is more common to stir empathetic feelings in viewers, suspenseful and "loud," hardcore themes are consistent with shock-building formulas in these types of films, as they are used to emphasize ideas of conspiracy and secrecy, or to aggravate feelings of rage, respectively. Shock-based animal advocacy documentaries often skillfully add sound editing that measures out the film's rhythm and emotional build-up, as images of violence are accompanied by intradiegetic nonhuman animal sounds (usually of terror or distress) and intradiegetic noise of the machines and instruments employed in farming plants and other spaces of exploitation. Beyond soundscapes, other factors, as we will see, such as the visual focus on the material body and its texture or the evocation of excess and saturation, immerse the viewer in graphic representations of violence. And it is important to understand that the *raison d'être* of this violence is to produce moral shock.

Moral shock, James M. Jasper and Jane D. Poulsen explain, occurs "when an event or situation raises such a sense of outrage in people that they become inclined toward political action, even in the absence of a network of contacts" (1995, 498). Although these situations are usually "public events, unexpected and highly publicized," they can also materialize as "experiences of the individual." They go on to add that "organizers try hard to generate [moral shocks] through their rhetorical appeals," the most effective being those that are "embodied in, or translatable into, powerful condensing symbols" (498). In this manner, moral shock is defined as a calculated strategy used to raise awareness, foment emotional engagement and recruit sympathizers that will seek out the political and/or communal means to denounce a given state of affairs. The types of symbols or indexes that can trigger shock may vary. What makes the rhetoric greatly effective is the symbol's capacity to polarize the identity of the agents involved, presenting a dichotomous worldview that separates perpetrators from victims, heroes from villains (Jasper 2018, 149) – a classification that brings order and meaning to the moral dilemma that the subject cannot unsee or in good conscience turn their back on. Following this idea, Jasper associates moral shock with affective responses, localized as "extremely bad moods" that may build up to a somatic reaction: "The shock originates in reflex emotions that arise when we discover that the world is not as it seemed, and we are startled, stunned, sad, angry, disgusted and/or disappointed" (2018, 87).

In animal advocacy and activism, discussions on the ethics of moral shock tactics are bound to their effectiveness as strategies that can precipitate transformative change. Laura Fernández's survey of the response to moral shock among European activists concluded that the tactic of exposing viewers to "the graphic, violent representations of nonhuman animals in situations of exploitation and slaughter" (2021, 139) works best when combined with additional visual strategies that emphasize adjacent matters such as nonhuman animal individuality and subjectivity, or infographics about industrial practices (153). Fernández's research indicates that *Earthlings* ranked as the documentary with the highest impact on the activist interviewees (143), not unlike Huw D. Jones's recent study on the affective response of European viewers to documentary film as a genre, which singled out *Earthlings* as a film to which survey respondents had been particularly receptive (2022, 194, 195). Additional studies such as Annie Bernatchez's (2024) have also insisted on the emotional toll that animal activists face and the traumatizing effect of not only visuals but also smellscapes, soundscapes and touchscapes. This goes to show the extent to which the corporeal and the somatic are central to moral shock and to what Niklas Hansson and Kerstin Jacobsson (2014) term "micro-shock," that is, activists' re-motivation and reinforcement of a movement's moral convictions through voluntary (re)exposure to distressing material.

Beyond issues of the efficacy of moral shock in general and of animal advocacy documentaries in particular, an examination of the instigators (i.e., Jasper and Poulsen's "powerful condensing symbols") is essential to understand how emotions are triggered, blunted and somatized in such a way that filmic (audio)visual indexes become corporeally assimilated and/or rejected. As will be discussed later on, *Earthlings*, *Lucent* and *Dominion* immerse audiences in deeply disturbing images of nonhuman animals being neglected, violated, hit, prodded, broken, skinned, boiled, sliced, hung, bled, whipped, thrown, stepped on, kicked, burnt, twisted, thrashed, gassed, battered, smashed, electrocuted, drowned, stabbed, trapped, ground and thrust by humans and the industrial machinery and equipment at their disposal. They are also humiliated and verbally abused by breeders, trainers, scientists or frustrated workers at their stations in the production line.

It is this brutality unleashed upon visually terrified bodies, and the synesthetic plasticity of how such bodies are physically tormented, that qualifies these physical actions as a more "traditional" type of violence. These representations contrast with what Rob Nixon (2011) terms "slow violence," where graphicness is supplanted by insidious, gradual (and even invisible) forms of assault that require "a more active pursuit as the effects tend to be hidden from view" (Chang 2016, 108). I will return to the notion of slow violence and its association with utopic and dystopic visions of the Anthropocene in Chapter 6; for now, suffice it to point out that the type of violence depicted in this chapter draws on explicitness and graphicness without shying away from detail for the purpose of moral shock. Violence simultaneously becomes too much (the continuous and seemingly endless piling of extremely distressing footage) and not enough (the documentaries, as anthological and archival products, suggest that they only show a portion of the continuous and routine abuse that nonhuman animals experience every day in the animal-industrial complex).

Unsurprisingly, the signature undercover footage that is the bulk of these films, especially that retrieved from factory and fur farms, has often been connected to the horror genre for its brusque array of bloodbaths, twitching bodies, cold, dark, dirty spaces and terrified screams, among other motifs of horror film's lore. Sarah O'Brien reminds us that classic horror movies such as *The Exorcist* (1973) and *The Texas Chainsaw Massacre* (1974) "mix sounds sourced from slaughterhouses to heighten thrilling scenes of carnage" (2012, 124). Precisely because of their availability as a resource to excite viewers of fictional films, images of nonhuman animal slaughter "traverse boundaries between fiction and documentary, art-horror and natural horror" (Middleton 2015, 286), which tells us something about how their suffering and mutilation can be commodified and aestheticized in morally questionable ways.

The violence that is at the center of these types of animal advocacy films is in some ways also reminiscent of other types of shockumentaries, such as the mondo films that began to capitalize on racist, speciesist and sexist sensationalism in the 1960s. Yet the attempt to push viewers toward transformative action precludes the type of aestheticizing of violence that had characterized the mondo subgenre. Most famously beginning with the Italian film *Mondo Cane* (Jacopetti et al. 1962), which set out to record, anthologize and ironize about "savage" customs across the world, mondo productions displayed a penchant for the voyeuristic from an ethno- and Eurocentric gaze. *Mondo Cane*'s many scenes of graphic and slow violence against nonhuman animals are treated by the camera and narrator as objects of fascination in themselves – historic records that attest to the existence of atavistic rituals and practices and that test the artifice of modernity and progress. In addition, *Mondo Cane* came "at a precise time when the short newsreel became a cinema 'larger than life' and perfectly exemplified and exploited a growing fascination with such 'knowing' through the big screen" (Goodall 2006, 120). The perverse politics of mondo films, therefore, were enabled by new epistemologies in journalism and film that sought to visually colonize the racial/sexual/species other as part of their sublimation of feelings of shock. Their parading of subjects, transformed by the camera into "attractions" (Nichols 2001, 87), should not be read reductively or simplistically (Goodall 2006), and has been associated with a predominantly teenage male audience (Brottman 2004, 68). Most importantly for this discussion, mondo films' concern with the entertainment and commercial value of an aestheticized, historical record of violence, is opposite to the transformative and political agenda of advocacy and activist nonfiction film.

As indicated in the introduction to this book, the films created in the 1980s and 1990s by animal rights organizations set certain standards in terms of raw material, montage and narrativity to which later generations of activists and filmmakers would return. In fact, it is not uncommon for undercover and stock footage of these films to be patched into more recent documentaries. Many of these earlier films focused more exclusively on vivisection and animal experimentation, and crafted representations of the scientific community that echoed the nineteenth-century humane movement's perception of the medical researcher as an unscrupulous monster driven by blind male ambition and bloodthirst – an image that made its way into popular imagination with the publications and pamphlets distributed by British and American anti-vivisection societies and penned by notable figures such as Frances Power Cobbe, Ouida, Albert Leffingwell or Mark Twain, among others.[2]

Unnecessary Fuss (Newkirk and Pacheco 1984) reached a wide audience thanks to the distribution of copies by People for the Ethical Treatment of Animals (PETA) to the *New York Times*, the *Washington Post*, NBC's "Nightly News" and CNN, and the facilitation of two screenings on Capitol Hill (Lowe 2006, 185–86). The film not only showed repeated violations of the Federal Animal Welfare Act by scientists of the University of Pennsylvania's Head Injury Clinic but also, in an effort to trim nearly sixty hours of undercover footage retrieved by the Animal Liberation Front (ALF) into a film of approximately thirty minutes, foregrounded the image of the laboratory as a place of horror, where baboons were subjected to inhumane and torturous blows on the head to cause brain damage, and were mocked and insulted by researchers indifferent to their pain and suffering. This filmic framing of the laboratory space was heavily criticized by defenders of animal experimentation. The veterinarian and neuroscientist Adrian R. Morrison complained about being misquoted in the film, and dismissed the misbehavior of "young, laboratory workers" as a somewhat understandable act of "psychological relief" – their "humor" could be rationally explained as a defense mechanism with "curative power," and was (Morrison conveyed) unfairly "exploited" in *Unnecessary Fuss* (2009, 34). Although Morrison did admit to some of the clinic's faults, he did so in a minor way, softening the scientists' (moral) responsibility. Morrison cleansed their "humor" by reinstating an elitist view of the scientific community that asks that researchers not be judged by the same ethical standards as common citizens. The pressures of scientific experimentation, it is inferred, must excuse (at least to a certain extent) researchers' coping and socializing behaviors in the laboratory. Morrison's criticism of the film fomented accusations against perceived narrative inaccuracies. His stance was part of a response of a wider community that characterized the film's editing as tendentious and bigoted, leading to a misrepresentation of the clinic's regular procedures. Filmic predisposition toward storytelling facilitated the condensation of footage into a synecdoche where a part represented the whole, and this was supported by a narrator that could authoritatively devise what these critics perceived as a misconstrued or false explanation of the visual context. These manipulations, they believed, misled viewers toward needless moral and ethical questions. Public outrage was, of course, what PETA sought, and the very title of the film was stylized as a pun that exposed the clinic's unscrupulousness:

the head of the clinic, Thomas Gennarelli, had previously stated in a newspaper that he refused to publicly discuss laboratory research so as not "to stir up all sorts of unnecessary fuss among those who are sensitive to these sorts of things" (quoted in Finsen and Finsen 1994, 68).

As antecessors of today's animal advocacy documentaries, *Unnecessary Fuss* and other short films in this line greatly contributed to the association between the animal-industrial complex and aesthetics of violence and horror. At the same time, there was a progressive prototyping of the textures of hidden camera recording (the shakiness of handheld devices, poor high or low angles that often preclude a clear picture of the action and rest on narrators' explanations for support, night vision, often granular or fuzzy resolution etc.) that gradually established themselves as a cinematographic insignia of authenticity (Freeman and Tulloch 2013, 114). Poor technical and visual quality was directly proportional to the premise that exploitation and institutionalized violence against nonhuman animals happens in secrecy, far from the sight of citizens who would be appalled if they knew what went on in these spaces. Being appalled is not enough, however, and moral shock can only be successfully completed if the "extremely bad moods" (Jasper 2018, 87) trigger the individual or the community enough to resort to action. Perhaps one of the miscalculations of the animal rights and liberation movement has been to overstate and overestimate the dictum that to bear witness will lead to moral shock.

Documentaries have not just resorted to form and cinematography to authenticate secrecy; some have integrated the motif of the hidden into their plot structure. Jenny Stein's *The Witness* (2000) revolves around a New York builder's single-handed activism against the fur industry, climaxing with his driving around the streets during the Christmas holidays in a white van customized with a large screen on the side showing undercover footage of cruel industry practices. The film succeeds in exposing the fur industry by showing the images, but is arguably more invested in portraying people's first response to them. As a meta-film, we watch a documentary in which passersby watch filmed images of animal cruelty: while in the background we catch sight of walkers glimpsing but moving along, Stein's frontal focus is on the faces of the people who have stopped by the van and are visibly stunned, moved or horrified by what they are beholding. There is some respite brought on by the slow motion, the intimacy of the night and the compelling soundtrack of Sarah McLachlan's "Angel." In this emotional culmination, the film appeals to viewers' empathy toward nonhuman animals by focusing on human intermediaries of the meta-structure: as the passersby become witnesses, they channel and give physiognomic visibility to the response that the documentary viewer will identify as the moral one. As Stephen Prince reminds us, film has the ability

> to capture the subtleties and nuances of socially resonant streams of kinesic expressions . . . via close-ups and other expressive devices, to intensify and emphasize the most salient cues for the viewer's understanding in cognitive and affective terms of the meaning of the scenes depicted on the screen.
> (1993, 24)

In *The Witness*, the recognizability of human physiognomy facilitates the indexical correlation between gestural response and the violence triggering such reactions.

Deploying again the motif of conversion-through-visual-exposure is PETA's tellingly titled short video *Glass Walls* (2009), featuring Paul McCartney, which begins and ends with the quote that "if slaughterhouses had glass walls, everyone would be vegetarian." The video contains a stockpile of images of copious violence perpetrated in the animal agriculture industry, including fishing. As a probable response to PETA's own penchant for quoting adversaries, in the early 2010s the North American Meat Institute (NAMI) began releasing installments of short online videos as part of its own Glass Walls Project (Meat Institute 2012–2018). With renowned animal scientist and welfarist Dr. Temple Grandin as host and trusted tour guide, the videos, which followed the steps of different processing plants, were strategized as a calculated exercise of transparency to cleanse the reputation of the meat industry. Far from being represented as spaces of gore and horror, the plants were aestheticized as clean, safe places where nonhuman animals were humanely treated with calm handling that reduced stress and fear. Humane treatment, the videos showed, involved the practice of techniques and dynamics according to a species' demonstrable group behavior and instincts. This framing fits the type of propagandistic discourse that Olatz Aranceta-Reboredo and Núria Almiron classify as "fake compassion," because behind an "ethical façade" (2023, 40) there remains a reinforcement of speciesist ideologies and practices. With their ethics-washing, the videos devised "everybody wins" scenarios where animal welfare is presented as humane and cost-effective, and guarantees quality meat. When compared, PETA's *Glass Walls* and the NAMI videos provide just another example of polarized forms of spatial representation in the history of documentary, where the former aims at moral shock and the latter at social appeasement and industrial reassurance. Both rhetorically insist on the notion of transparency to reinforce their semiotics of iconicity of the "truth," where what is narrated must be consumed as an absolute equivalence of the profilmic scene. At the same time, nonhuman animals are indexed into correlative meanings, where an uncontrollably trembling body, a lowered head and upwards-glancing eyes might point to fear and suffering; or, in the case of the NAMI videos, a group of sheep following one another to the kill floor signifies calmness and natural behavior, thus emblematizing the virtues of welfarism. As a tactic of manufactured consent to the exploitation of nonhuman others, rather than have the meat industry shrink further into the unseen, the NAMI videos create the illusion that they are delivering just what skeptics and activists have asked for: access and penetrability. Implicitly, they are dismissing the concerns of animal liberationists as paranoid and unfounded and reducing black-bloc trespassers to self-deluded felons.

In their grammar of reality and truth, both the PETA and the NAMI videos make the mistaken assumption (or at least, perhaps driven by propagandistic methods, they *seek to appear* to assume) that spectators will, respectively, either adopt a vegan lifestyle or consume meat with a guilt-free conscience. The postulation of the viewer as *tabula rasa* and as a speck in a uniform mass of unsuspecting consumers of (audio)visual images is, in the view of Carl Plantinga, characteristic of an outdated, post-structuralist frame of mind that "*cannot* recognize differences between spectators" because it "build[s] the ideological effect of the cinema *into the apparatus itself*, proclaiming it a kind of universal effects machine that works on everyone the same way" (1996, 318; emphasis in original). Plantinga's call for what he termed the "instrumentalist approach," conversely, demanded at the end of the twentieth century a more scrupulous study of the "diversity of uses of nonfiction film" with the potential to recognize "a diversity of spectators and a heterogeneity of responses" (1996, 318). Of course, reception studies and response theory have come a long way since then, and the dismissal of audiences as universally credulous by film theorists has long been overcome. However, the confrontational attitudes between animalists and animal exploitation industries that continue to escalate and spread in other (digital) media (Stokes and Atkins-Sayre 2018) often cannot, because of the polarized conceptions of morality they operate by, seem to shed a tacit understanding of viewers as overall malleable and persuadable.

Specifically, on the side of animal advocacy and contrary to such strategies, it is not uncommon to find literature on the spectrum of responses that both activists and non-activists can experience after having been exposed to shocking images. To get back to Jasper's notion of the "condensing symbols" and their triggering of "extremely bad moods," responses to shock do not necessarily lead to political action but can cause paralysis instead: "We enter a disoriented mood, which may eventuate in our rethinking or reasserting our basic moral commitments, in action to set things right, or alternatively in depressed withdrawal" (Jasper 2018, 87). Such withdrawal may range anywhere from numbness, outright rejection and conscious or unconscious efforts to lock away the traumatic experience, to the often-discussed shared feelings of burnout and compassion fatigue. As a defense mechanism with proactive aims, activists frequently turn to strategies of coping, desensitization and disengagement from shock that can make their commitment more sustainable in the long run (Wrenn 2023).

And then there is also the possibility that violence may not prompt any shock at all, and that viewers may feel unaffected and indifferent to nonhuman animal representations of suffering. As animal photojournalist Jo-Anne McArthur, who has dedicated her life to uncovering the cruelty behind animal industry practices, explains:

> You don't know who the person that you're trying to reach with an image is. Some people have had a lot of exposure to sadness and are numb to your image, or shut down to it and they're not open to it. . . . Some people are really receptive to seeing an image and then being told to go vegan. . . . But it's generally not like that. People need contact, and time, and compassion. (2023, 144)

For one thing, visual overexposure and discourses of manufactured consent can lead to insensitivity and the impediment of an empathetic encounter. While activists and advocates may be receptive to the slow violence exercised in zoos and aquaria, these signs may not necessarily be perceived by average spectators. And while much of the advocacy against animal agriculture, the fur and the medical research industries gravitates, as suggested thus far, around ideas of secrecy and concealment, let us not forget that the entertainment industry often involves very graphic acts of traditional violence for show purposes that are condoned by different world cultures and that still draw in large crowds, thus challenging *The Witness*'s and *Glass Walls*' inference that one will change once one sees.

Approaches and Perceptions

In this section, I examine how graphic violence against nonhuman animals translates into the stylistics of *Earthlings* (2005), *Lucent* (2014) and Dominion (2018), all of which purposely aim at moral shock by way of visual and aural exposure to animal cruelty.

Shaun Monson's *Earthlings* can loosely be credited as the film that, perhaps propelled by the public's reawakened taste for documentaries at the beginning of the twenty-first century (Rich 2006; Smaill 2016, 2), defined an aesthetic that was consistent with a growing uneasiness about planetary impact and that implicitly tested the tolerance limits of post-9/11 audiences for violence and nonfictional visual horror. Coherence and unity are also imprinted in its structure, as viewers are guided through five industry categories (pets, food, clothes, entertainment, science), with each part conceived as a repository of violent moving images of what are described as routine practices in such industries. Backed by the authoritative weight of two vegan celebrities (Joaquin Phoenix in his first performance as a narrator in an animal advocacy feature-length film and Moby as musical composer; see Chapter 5), Monson and his production company Nation Earth delivered more than ninety minutes of footage and hidden camera recordings of animal cruelty, torture and abuse.

A predecessor of this form of cataloguing is arguably Victor Schonfeld and Myriam Alaux's *The Animals Film* (1981), a feature-length documentary narrated by Julie Christie that traversed the inside of animal exploitation industries. Stephen Marcus Finn calls *Earthlings* and *The Animals Film*, along with Delforce's *Dominion*, "omnibus documentaries" (2023, 75–88) on account of the magnitude of their scope. As he notes as well, *Earthlings* borrows from *The Animals Film* some anti-speciesist arguments and stylistic choices on the work of visuals in relation to sounds (82). However, *Earthlings*, I would argue, noticeably resorts more to the performative mode than the more consistently expository *The Animals Film* because of its reliance on musical eloquence and its sequencing of images. This is a tricky statement because there remain some performative elements in *The Animals Film* and because Nichols characterizes the performative mode as one that downplays objectivity in favor of the subjective experience (2001, 34). *Earthlings* certainly does not question speciesism as an objective reality (nor will most advocacy documentaries, because such reality is what motivates advocacy in the first place); however, it prescribes an epistemology of such reality through the affective. Knowledge, in this way, becomes "concrete and embodied" where "experience and memory, emotional involvement, questions of value and belief, commitment and principle" (Nichols 2001, 131) influence our understanding of the world. While *The Animals Film* frequently cuts into interviewees and people discussing the abuse of nonhuman animals (offering points of views from scholars, activists, farmers, scientists, or random people on the street) and relaxes the tension with interludes of government films, newsreels and cartoons, and by zooming in on the small print of product advertisements, *Earthlings*' texture is more exclusively centralized on the look and sound of violence, thus sanitizing the narrative from interventions that may interrupt the emotional and affective saturation it aims for. More importantly, like in the rest of the 1980s animal advocacy documentaries, the visual quality of *The Animals Film* has aged in a way that could soothe the viewer into activating a historical lens. From a twenty-first-century perspective, the faded chromatism and the comparatively fuzzier resolution, adding to the 1970s and early 1980s fashion donned by the multiple people shown, anchor and historicize the violence, endangering the exegesis by becoming "too archival," that is, by creating the false perception that "this happened long ago" and that, therefore, it is acceptable to favor a historical and historiographic approach to the material over an affective one. As in the case of other long-standing conflicts, animal advocates find themselves under the obligation to renew and restore images to "freshen up" and (re)activate ethical commitment on the part of viewers. *Earthlings*' supply of images does, admittedly, sometimes borrow footage and material from earlier films and recordings, but it succeeds in transcending the historical mindset by integrating them into a wider audiovisual discourse that resets the timer on its own present (the beginning of the twenty-first century).

Returning to its structure, it should be noted that the montage of *Earthlings* serves a didactic agenda different from the type of approach employed by earlier documentaries reliant on shock tactics (aside from *The Animals Film*). PETA and other organizations had, throughout the 1980s and 1990s, developed a style of filmmaking that, as we have seen with *Unnecessary Fuss*, built its persuasiveness around information that was strategically condensed to fit into a short format targeting specific centers and companies. These localized approaches strengthened the legitimacy of the films' claims. The activists had infiltrated these spaces and had dedicated their time to doing their research, and this dedication contributed to the authoritative weight of their findings. By targeting specific companies and drawing as much media attention to them, activists could more effectively pressure industry agents to change their policies and regulations. They could direct public frustration and outrage against an identifiable business or institute and could therefore more easily channel action toward achievable goals (e.g., the boycotting or lobbying against that particular target), and they could perhaps compile enough evidence to urge funding bodies to cease their financing or collaboration.[3] *Earthlings*, on the other hand, represents speciesism as an epidemic, a global evil that is simultaneously supra-cultural (it is widespread throughout different lands, continents and oceans) and culture-specific (with footage of practices carried out in specific nations), and it connects eastern and western geographies via a capitalist market of nonhuman animal products and by-products that are traded and circulated across the world.

This notion of interconnectedness subverts what Jonathan Burt once loosely termed the "slaughterhouse aesthetic," by which he referred to a type of dispassionate camera recording of industrial processing of nonhuman animals that reproduces an image of mechanization and anonymity, and that fails to connect these manufacturing processes of repetition with the wider cultural mechanisms of (capitalist) production (2002, 174–75). Sarah O'Brien (2012, 108–10; 2023, 45–75) refers to the slaughterhouse aesthetic as a means that maximizes visibility in a way that leads to invisibility: the hypervisibility of what is usually concealed (the trope of the hidden that was discussed in the previous section) enables the visual recognition of the horrors of the meat industry, yet the shocking effects of such violence and the hypnotics of the mechanical processes of industrial order and copy, isolate these practices away from their socioeconomic and cultural embedding: "[I]n its efforts to show slaughter to the exclusion of all else," O'Brien writes, "the slaughterhouse aesthetic renders these connections invisible" (2023, 53). Where Burt describes Georges Franju's *Le sang des bêtes* (1949) as a film that evades the slaughterhouse aesthetic because of its visual symmetrizing of the abattoir and the historical inflections of Parisian urbanism (2002, 175–76), I contend that *Earthlings*, which wields "make the connection" as its motto, succeeds with a stylistic form that contests a separation between the slaughterhouse (or, to that effect, the rest of the industrial spaces it covers) and the consumerist markets that sustain and justify their existence. It is true that the way in which these connections are addressed in the film is not through in-depth discussions or the compiling of evidence and information typical of expository and investigative journalism, but rather through a script that sticks to a threaded description of the production process, in which the semantic field associated with sentience is paired with disturbing images and sounds of pain and suffering. In this manner, Monson traces production and animal rendering across cultures, and by amalgamating similarly styled representations of each industry, he conjoins them under the single theme of "speciesism with variations."

Adding to the motif of a connected, planetary web of exploitation, are *Earthlings*' links with other forms of abusive power based on hierarchical binarisms that elevate a normative category over an "other." Drawing on the equation proclaimed by animal rights scholars and activists since the last quarter of the twentieth century (Singer [1975] 1990, i–vii; Spiegel 1988; Adams 1990, 2003; Dunayer 1995; Patterson 2002; to name a few of the most popular publications), *Earthlings* (like *The Animals Film* before it) insists on the correlation between speciesism, sexism and racism as forms of institutionalized and oppressive power relations (see Chapters 3 and 4 for a more detailed discussion of these analogies). As part of this rhetoric, Monson maximizes the semiotic potential of the term "earthling" to convey a cross-species sense of self-awareness and subjectivity shared by all sentient beings under the triad of "nature, animals and humankind."

It allegedly took Monson six years to compile the moving images collected in *Earthlings*. The film brought together his own recordings of animal shelters with stock footage and the undercover work of organizations such as PETA, Last

Chance for Animals (LCA), The Humane Society of the United States (HSUS), Stop Huntingdon Animal Cruelty (SHAC), the Humane Farmers Association, Farm Sanctuary, Sea Shepherd Conservation Society and Greenpeace. As an artisan of montage, therefore, Monson's originality as a filmmaker greatly rested on his vision for meaningful arrangement of previous material that could make the most of its compositional and textural potential – an architectural nod to the post-structuralist "death of the author" and his reconfiguring as "scriptor," whereupon critical engagement with reader/audience response acquires a more central significance than the figure of the writer/filmmaker (Barthes 1977, 142–48). Construed as an activist take on the "text-as-intertext," *Earthlings* recapped, through form, on the cumulative. Monson described the ambition behind the film as follows: "I thought, someone's gotta do this. Someone's gotta put this thing together, like the encyclopedia film, the ultimate encyclopedia movie of how humans use animals" (Monson 2019). The anthological breadth certainly garnered critical attention. *Earthlings*' 2005 accolades included Best Animal Advocacy Documentary Feature in the Artivist Film Festival, Best Content Award at the Boston International Film Festival and Best Documentary Film and Humanitarian Award (for Joaquin Phoenix) at the San Diego Film Festival. Amassing moving images of suffering not only illustrated the global spread of speciesism but also established the archival functions of the film, creating a "museum of present horrors" of sorts. Time will tell if, like *The Animals Film*, *Earthlings* will become "too archival," not only because the caducity of digital quality foments a historiographic approach but also because documentaries' ties with the present are delicate and unstable. John Blewitt (2010, 100) claims that films about current issues do not enjoy much commercial longevity. Considering that advocacy films, intentionally and by their very nature, focus on the present as the temporal axis that demands transformative change, their appeal to wider audiences is always precarious.

Chris Delforce's *Lucent* is tactically dissimilar to *Earthlings*, as it recovers the strategy of targeting one specific industry (pig farming) in a more localized space (Australia). Delforce founded the Farm Transparency Project – formerly called Aussie Farms – in the early 2010s to tackle the issue of animal agriculture in Australia. As suggested earlier, these approaches can inspire more manageable actions on the part of shocked spectators: having a close, identifiable target allows a workable sizing up of the situation (as opposed to an adversary considered too large and pervasive, which can lead viewers to feelings of helplessness and personal insignificance). The film is the result of two years of handheld and hidden camera recording of standard practices in over fifty Australian piggeries and slaughterhouses, with a few final minutes dedicated to other nonhuman animals that are farmed for food.

Four years later, in *Dominion*, Delforce attempts a more holistic scope, and at the beginning of the film viewers are instructed to regard the Australian bulk of the material as representative of practices across the West, moving from the endemic to the epidemic inertia of speciesism. *Dominion*, which features Monson as co-producer, soon catches up with a revised structure of *Earthlings*, as it encyclopedically segments itself into categories – this time, according to species. Starting with the pigs that were heavily documented in *Lucent*, it moves on to egg-laying hens, broiler (meat) chickens, turkeys, ducks, cows, sheep, goats, fish, rabbits, minks, foxes, dogs, horses, camels, mice, exotic animals, and seals and dolphins. Violence runs through several sectors, and, as in *Earthlings*, Delforce confronts the meat, clothing, entertainment, sports and research industries. In the same way too, it uses celebrity narrators (with Phoenix once again as leading voice, and a star-studded cast including Rooney Mara, Sia, Kat Von D and Sadie Sink). Furthermore, the parallels with human slavery and the Holocaust are emphasized with historical footage and photos. *Dominion* went on to receive numerous awards in 2018, including Best Director at the Los Angeles Independent Film Festival Awards, several Awards of Excellence at the Hollywood International Independent Documentary Awards, and Best Director and Best Documentary Feature at the Festigious Film Festival.

The camera work in Delforce's films is much cleaner and clearer than in *Earthlings*. The temporal and technological gap is not as wide as between *The Animals Film* and *Earthlings* (which favors today's perceptions of the latter as an abiding testimony to the present), although certain nuances are still palpable. The digital technology of more modern equipment, whether handheld or hidden, delivers, for the most part, better-quality footage bringing luminosity to traditional shocking images, even in the darkness of factory farms. *Lucent*, after all, is not an anthological compilation of previous material, but the result of active research on the systemic violence practiced in Australian pig farming. In honor of its title, the film seeks to shed light on this industry and enlighten viewers along the way. If, as stated earlier, low-resolution, rushed and impossibly angled footage has been appropriated by animal advocates as a staple of authenticity, what *Lucent* delivers is a technologically updated version of these still recognizable features. Without sacrificing the essence of this hallmark of activist filmmaking, Delforce delivers a representation that is as transparent and tangible as possible for legitimation purposes. *Lucent* and *Dominion* also provide some insight into activist undertakings, from suspenseful depictions of the placing of cameras and trespassing into facilities to the gentle treatment of the victims – a kind caress, a hand helping a pig pull her leg out from a crack on the floor, or just bringing water to an exhausted sow who cannot lift herself up. These instill a sense of pathos that cannot help but glorify and ennoble their cause. There is also a patience to the rhythm (more noticeable in *Lucent* than in *Dominion*) that feeds viewers' own growing claustrophobic desperation of being trapped in these spaces of horror. This cramming of extreme emotions contributes to the moral shock tactic that I will later describe as the principle of quantity.

With their compilation of violent moving images, Monson and Delforce implicitly ironize about viewers' role as passive consumers in capitalist systems. Hyperrealism and consistent, systematized violence abruptly shake viewers from their comfort zones, and for some, the unbearableness of watching is assimilated as a means of purging and castigation of the self. It may be the case that the shame brought on by the recognition of one's own participation in the perpetuation of violence (namely through the consumption of animal products and by-products) is deemed more acceptable than the shame brought on by the deliberate choice to look away. In relation to what qualifies as moral and immoral responses, Elisa Aaltola (2014) discusses the ethical problematics associated with images of suffering of nonhuman others, asking how these can be actively and persuasively communicated. Drawing on Susan Sontag's *Regarding the Pain of Others* (2004), she recapitulates on the distinction between "normative" and "aesthetic" forms of looking at suffering – the distinction being based on whether one is inspired to do something for the victim or whether the response is one of inaction, which risks leading to emotions of indifference at best or pleasure at worst (what in German is termed *Schadenfreude*). In other words, moral shock can only be moral if one is "to act on behalf of the victim. As long as images of suffering raise the desire to do something in order to help the sufferer, they should be produced and looked at" (Aaltola 2014, 25). Aaltola effectively digs into the many complications that the act of looking at nonfictional suffering raises and addresses the victims' rights to privacy and the viewer's right to not be exposed to such violence. Like Sontag, she concludes that the discomfort and sense of violation caused by one's perceived right to not look does not justify not looking, except in the case of people who are already sensitized and are already "doing" something for the victims (this brings us back to activists' reignition of their strength and perseverance through micro-shocking). What Aaltola does emphasize is the necessity to accompany images with viable and manageable directions for action, given the very real threat of detachment caused by aesthetic perception and the laws tightening their grip on the exposure of animal agriculture practices.[4] Aside from boycotting products, going vegan, or ceasing visits to zoos, circuses and aquaria, other fundamental actions suggested by activist films include future contributions to (crowd)funding and the screening and showing of the films in order to keep the images alive and running in media fora and digital geographies. Additionally, Delforce's representations of caretaking and affection toward victims by activists is intended as a model to support the normative approach. These portrayals insist on an ethos of kindness that can be applied in other spaces of interspecies interaction that viewers are more likely to occupy, and they propose models of small but meaningful acts of compassion to engage with nonhuman animals. The possibility of achievability is, again, important to sustain action in the long run.

Yet these responses also beg the question: are these actions sufficient to classify how one looks at suffering nonhuman animals as fully normative? Is there a scale, a degree to normative approaches? Are there "more normative" and "less normative" approaches on account of how people channel their shock into action? Aaltola does not refer to such possibilities, which may lead to the inference that any action, small as it may be, accounts for a normative approach, including the mere response to the image as one of compassion and care. This is a fair interpretation; after all, every little bit helps in an activist cause, and it is detrimental for the movement to have viewers thinking that their actions do not matter in the grand scheme of things; that they can make no difference whatsoever. For theoretical purposes, however, the question is of consequence because it demands that we revisit and pry open the idea of "doing something for the victim." Except for a handful of nonhuman individuals that the activists remove from farms and take to sanctuaries, the rest of the victims circulating through the assembly lines and manufacturing processes and populating Monson's and Delforce's filmic grammar are long dead. Thus, "to do something for the victim" does not actually mean doing something for that one particular individual or group of individuals; rather, the filmed nonhuman others become martyrs for the countless individuals like them to follow. As Aaltola writes, "[i]mages of animal suffering can form an integral part of a moral relation to *other* animals, and as such become a necessity rather than a nuisance" (2014, 25; emphasis added). But to repetitively watch or spread images of their suffering and death does not liberate them from their torment: our helplessly anthropocentric morality leads us to believe that we can do right by them by archiving their pain and death into a grander web of similar images of violence whose weight will eventually be enough to end animal exploitation industries. Moral shock, in other words, has a way of dealing with impotence that does not necessarily lead toward inaction, but seeks ways out to ease one's conscience by minimizing guilt. We fill our actions with meaning and purpose, an ultimate contribution to the rewriting of human-nonhuman animal histories where exploitation is not to be hidden or welfare-washed. We create a discourse based on honoring the memory of the billions of nonhuman animals who lose their lives in these industries – discourses which in reality cannot do anything for the nonhuman animals that were once filmed, whose primary interest was to live. These observations are not intended to direct, by any means, criticism toward this coping strategy, which is only natural and indispensable for the planning of a future advocacy agenda, but rather to illustrate why activists themselves still experience Post-Traumatic Stress Disorder (PTSD) or compassion fatigue despite their dedication and fruitful commitment. Any activist documentary where moral shock is bound to the certainty of the victims' death, navigates difficult terrain in terms of engaging viewers with a fully normative approach. Death is too finite and inescapable a place to offer a fully convincing sense of solace in one's moral responses and actions. It is impossible, therefore, to calculate the range of responses that audiences may feel toward shocking images of violence; we can consider their effectiveness on the grounds of whether transformative action is pursued or not, but the actual emotional circuits remain more obscure and challenge the filmmakers' ideal response on the part of viewers (Kääpä 2013, 109).

On the Body and the Bodies

Beyond the question of what to ethically make of the act of looking, animal advocacy documentaries reliant on moral shock also demand that we pay attention to issues of quality and quantity with regard to representation. The former addresses the strategies employed in the representation of nonhuman animal subjectivity using what is perceptible: their bodies. The latter builds upon such representations numerically, exploring the rhetorical potential of repetition and plurality. In the case of Monson and Delforce, this is strongly bound to their films' holistic, structural ambition. Tal Morse (2021) discusses how the media, as a moralizing agent, negotiates its place with regard to the representation of nonfictional death. The (in) visibility of death, he contends, involves issues such as voyeurism, spectacularism, the trauma of shock and the social conventions associated with taboos and mourning. "[T]here are various occasions in which death is more likely to be visible," he adds, for there are "variables like national identity and cultural geographical proximities [that] can explain the likelihood of showing explicit death images" (2021, 129–130). Although Morse and the host of scholars he cites are fundamentally concerned with the death of humans, the question of species proximity is key to understanding the multiple possibilities underlying moral shock and the representation of violence. In this vein, recent studies on eco-documentaries and animal advocacy documentaries have focused on the notion of the body as a site for empathetic experience and a trigger for somatic (self)recognition, thus overcoming its situatedness as just visible and visualizable matter (O'Brien 2012, 2023; Smaill 2014–15). These studies mostly draw on the work of Vivian Sobchack (2004), who theorizes about the subjective connections with on-screen characters through the recognition of bodily resemblance and the shared condition of mortality – a form of identification where ethical spectatorship is dependent upon the film's genre as fictional or nonfictional. Brought to the domain of the species discourse, Sobchack's theory is helpful to address how corporeality and the synesthetic rhetoric of animal advocacy documentaries accentuate interspecies similarities, and thus enable viewers to empathize with the subjectivity of nonhuman animals. This anagnorisis leads to an epistemological reconsideration of nonhuman animal bodies and is dependent upon representational choices in filmmaking: Smaill (2014–15), for instance, notes how representations of the sickly and moribund nonhuman animals that populate factory farms (and that disrupt pastoral images of traditional family farming) denote a somatic sense of repulsion and inedibility. O'Brien (2012, 121–31, 2023, 69–74) insists on the sounds of slaughter as unnerving indexes of corporeal identification.

The body as visible and audible matter can be further sensorially construed. Speaking from her own experience watching Jane Campion's *The Piano* (1993), Sobchack contends that films have the capacity to sensitize viewers at an epidermic level (2004, 61), and that prior to conscious thought, the cinematic encounter can first elicit a "carnal comprehension" (63). She further notes that:

> Although this experience of my body's prereflective but reflexive comprehension of the seen (and, hence, the scene) is in some respects extraordinary, it is also in most respects hardly exceptional. . . . [W]e do not experience any movie only through our eyes. We see and comprehend and feel films with our entire bodily being, informed by the full history and carnal knowledge of our acculturated sensorium.
>
> (Sobchack 2004, 63)

In their films, Monson and Delforce do not hold back from the significance of the squeals and cries of visibly terrified and abused nonhuman animals, and to this they add epistemologies based on the senses of touch and smell. Unhurried close-ups on open wounds, cuts, swellings, lacerations, fly- and maggot-infested infections, broken limbs and protuberances constitute synesthetic techniques that summon tactile recognition. How bodies are violently handled by workers and machines – what Claire Parkinson, in her excellent study on embodied visuality, terms the "technological touch" (2018, 63) – also transgresses our comprehension of tactility: the plasticity of flesh is made palpable with images of skinning and eviscerating, necks twist backwards when throats are slit, fur-bearing bodies become stiffened by anal electrocution, dogs drop into the heaviness of their bodies when they are "euthanized," live male chicks are ground in macerators, the rubbery skin of dolphins is scraped as vehicles drag them by their tail down pavements, and one may sense the painful breaking of bones when still conscious cows and pigs are hung from their delicate legs, or when chickens are grabbed and swung from their necks. The shock triggered by tactile recognition derives not only from images of nonhuman animals' tormented bodies but also from the visualization of the human body as an arsenal automated into violence. Routine movements at workstations and deliberate acts of cruelty (stomping, kicking, prodding, punching etc.) reconfigure the (predominantly male) human body as a weaponized instrument that restricts touch to violent interaction (as opposed to the nurturing caresses of activists in Delforce's films).

The smoothness of feathers, the softness of fur and the rough-haired skin of pigs is further violated by the cold wetness of urine and feces blended with the suppuration of blood and pus, situating the tangible body deep in insalubrious smellscapes of putridness where excrements and rotting corpses make the closed, dark spaces unbreathable. In this manner, the films' direct calls for empathy and compassion are intertwined with humans' necrophobic instincts to steer clear of putrid, dead matter. As Susana Monsó states, necrophobia is an evolutionary response strongly bound to the sense of smell, and a visual image of putrescence can trigger somatic feelings of disgust, even if the dead body does not represent a real danger for the onlooker (2021, 43).

The synesthetic quality of the images strengthens viewers' sense of identification through the phenomenological experience of what is sensorially perceptible and the subjective experience of suffering that is associated with embodiment. These somatic and affective connections reinforce the conception of the nonhuman animal body as factual, especially because the bodies that are represented are being harmed. Brutality, in other words, invigorates our proximity to the phenomenological accuracy of physical pain, and such accuracy attests to the nonhuman animal body as fact. This form of apprehending the other(ed) body, at the same time, befits documentaries' claims to truth and authenticity. As Nichols states, "[t]he sight of brutal violence, or its extreme manifestation, brutal death, engages us quite differently in documentary. This is not a simulation. . . . The imprint of history registers on the flesh" (1993, 190). Viewers' carnal recognition of pain and discomfort reaffirms the vulnerable body as a site that remembers, a place of memory. In the cinematic encounter produced by shock tactics, the shared experience of embodiment thus emerges as a reliable source of information and believability.

Monson's and Delforce's structural endeavors also demand that we pay critical attention to the saturating *number* of bodies. Notwithstanding their differences in rhythm and footage quality, *Earthlings*, *Lucent* and *Dominion* rely, as suggested thus far, on the technique of "more-ing," insofar as they cram image upon image of extreme suffering and overfeed it to an audience that is sickened and swelled by the hypervisibility of violence and cruelty. It is the principle of quantity as ethics, a quasi-bulimic trope that imagines the purging of our conscience as the vomiting and regurgitation of the stock images of bodies endlessly circulating through the production chain. Quantity becomes the twenty-first century: in an age of continual exposure to media violence, the quality of violence (i.e., the graphicness and spectacularism) can perhaps fall short of eliciting enough of an ethical response, or what Sontag and Aaltola would term a normative one. As a supplement, quantity does not necessarily seek to out-shock the previous image with heightened explicitness; instead, it authenticates violence and the believability of embodiment through systematized repetition.

There is a strategy to the dosing of violent images that by its own nature evinces an implicit belief in the potential of quantity as a moralizing force. Repetition of arguments and persuasive material in (documentary) film, of course, reinforces the chances of an emotional response in the audience (Plantinga 2009, 203; Terry 2020, 67). In other words, Monson's and Delforce's reliance on the trope of quantity is deliberate, to the point that it sometimes escalates beyond the portrayal of innumerable bodies and actively seeks to represent the epidemic scale of speciesism. Elevated shots come in handy for this: *Earthlings* begins with a view of the planet from outer space, and drone footage of factory farms is used in *Dominion*. *Lucent*'s opening with calm, pastoral images of pigs happily being themselves in a sanctuary is immediately cut off by a rapid succession of photographs of the miserable conditions of farmed pigs – a technique that is again repeated before the final credits. The effect of accumulation methodically reinforces negative responses to systemic exploitation, reflecting how every second of every day pigs are being abused and slaughtered. Furthermore, the duration of the films is also a nod to the principle of quantity as ethics. Anat Pick mentions that "[*Earthlings*'] collage of atrocities is a challenge to watch for a whole ninety minutes" (2013, 28). *Lucent* is fifteen minutes short of two hours, while *Dominion* reaches a two-hour mark. With the exception of *The Animals Film*, *Hidden Crimes* (Burgos 1986), and a handful of others, most animal advocacy films had, until the twenty-first century documentary revival, resorted to shorter formats. This is noted by Jonathan Burt (2005, 348) in his review of PETA's also conspicuously long *A Day in the Life of a Massachusetts Slaughterhouse* (1998). Monson and Delforce use extensive duration to visually reinstate the excess of harvested bodies and the magnitude of the issue.

In this way, the numbers of victimized bodies are instrumentalized to counter the discourses of quantity by animal exploitation industries and lobbyists. As capitalist structures, the mechanization of factory farming and animal agriculture appropriates the schema of cyclical production and accentuates it using lists of bodies copiously arranged not just in cost-effective ways, but also in visually methodical assortments that convey neatness and order. In *The Infinity of Lists*, Umberto Eco states that in the context of consumerist culture,

> [t]he technique of the list is not intended to cast doubt on any order in the world, on the contrary its purpose is to reiterate that the universe of abundance and consumption, available to all, represents the only model of ordered society.
>
> (2009, 353)

Pro-animal-exploitation aesthetics naturalize the lack of alternatives and manufacture consent to these forms of production through welfare-washing discourses (such as the one in the previously discussed NAMI videos) that appease public doubt and skepticism with their clean assortments. As a counterstrategy, *Earthlings*, *Dominion* and *Lucent* assimilate the schema into a rhetoric that seeks affective and somatic recognition, and to that effect they rework the motif of excess while lingering on the synesthetic repercussions of the indexes of violence that are engraved in those supposed spaces of neatness and order (which are in fact more accurately presented as dark, cold and filthy). Bodies, in other words, become re-signified as ominous indicators of chaos and moral confusion. Their excess in number does not visually comply with the packaged insignia of a capitalist assortment of abundance. The plurality of bodies does not instill any sense of composure. It is not just a matter of how these bodies look, but also of the tension between how they move (terrified, in pain, neurotically) and how they are immobilized (in crates, cages, or paralyzed by pain, fear and exhaustion, or even dead).

What does the idea of billions *look* like, after all? Is it even fathomable? Using shots of the many imaginative ways in which animal exploitation industries pack nonhuman animals in as little space as possible (rows of farrowing crates, grower pens or wire cages stored on top of each other, containing an indeterminate number of hens, foxes, minks, etc.), Monson and Delforce draw attention to the visual display of crowded bodies precisely to convey the impossibility of a headcount: numbers become countless in the most literal sense of the term. These types of shots are combined with a form of editing that reinforces the trope of quantity by leaning into sequencing. Even if the serialization is of images showing a lower number of bodies (or just individuals), it is the implied inertia of sequencing that suggests that the list could go on indefinitely. There are innumerable nonhuman animals beyond the frames of the shot, not just in those particular farms but in the world at large, and not just in the world as space but in the world as time. Billions of nonhuman animals have been killed before and billions will follow. There can be no order where and when there is no finitude to pain and suffering, only chaos. In this sense, the films comply with a type of "list" or "catalogue" that, in Eco's view, stands for "an *actual* infinity made up of objects that can perhaps be numbered but that we cannot number – we fear that their numeration (and enumeration) may never stop" (2009, 15). The entropy of billions of bodies (and the synecdochical filmic shots of collective and sequential suffering) saturates the screen, seeking to provoke moral shock.

It would be naïve, however, to ignore the possible negative side effects of the principle of quantity as ethics. The seemingly endless list and sequencing of bodies can also lead to numbness, impotence and indifference as a response, and activate the aesthetic form of looking at suffering described by Sontag and Aaltola. In the same way that one can gradually adjust to surpassing levels of graphic violence, a heavy dosage of image and body quantity can become palatable. The compositional similarity of the shots, with their repetition of indexes of pain and suffering, can also succumb to postmodern customization of images as a copy of a copy of a copy, and weaken the authenticity of the experiential singularity of each and all the beings that for a brief moment are the focus of these images. The recognition of individual subjectivity is at stake. It does not help that humans have difficulty in making out the precise physical features and physiognomy of individuals of other species. Face perception is associated with phylogeny, but environmental factors also have an impact (Berhmann and Avidan 2022). Personal contact or familiarity with a companion species or individuals from noncompanion species makes it easier for humans to physically and aurally recognize the subject; but this is hardly the case when our visual encounter involves groups of species that we have no actual, experiential contact with as live beings. A recent study on own-species bias suggests that considerable time is needed for humans to process nonhuman faces (Jakobsen et al. 2021), and time (by which I mean here sufficient duration of shot or sequence) is precisely what Monson's and Delforce's filmic grammar of compilation arguably lacks. In this sense, especially when it comes to dealing with factory farming, shots of countless bodies and body parts become amassed in our minds in a way that eliminates individual subjectivity in favor of a collective one. It is easy for the face of one cow to emblematize the face of all cows, or the face of one chicken to stand in for all chickens. And this hinders a proper historicization of that one cow or that one chicken.

Ultimately, of course, how viewers (un)consciously and carnally negotiate their cinematic encounter with nonhuman animals as groups and as individuals will depend on their previous knowledge, experience and exposure to violence. And it must be said that the inertia of sequencing, while on the one hand complicit with postmodernist de-authentication of images as copies of copies (which disrupts the principle of quantity as ethics), can be interrupted by an image that out-shocks anterior and posterior ones for the viewer. These images that are singularly compelling, break the cycle that threatens to lead to an aesthetic stance and become more memorable, pulling the viewer toward a normative response. Perhaps Monson and Delforce intentionally seek this effect by subtly nudging the viewer with a few more seconds of duration of shot, or by combining them with a specific musical or voice-over segment that appeals more explicitly to the performative mode. For example, the appalling close-up on a fully skinned fox slowly moving her head in agony in a Chinese fur farm (which appears in both *Earthlings* and *Dominion*), the distressing long shots of a kid, abandoned in a pen and desperately bleating for her mother (*Dominion*), or the focus on an anguishing sow whose back legs, the narrator informs, have become paralyzed after giving birth to a litter of stillborn piglets (*Lucent*), constitute haunting reaffirmations of individual suffering that demand moral action on the part of viewers. They are firm reminders that an ethical approach to visual multiplicity begins with an understanding of it being compounded by single, sentient subjects.

Conclusion

This chapter has dealt with the techniques and ethical implications of moral shock in animal advocacy documentaries that seek to maximize the exposure of the graphic violence that is systemic to exploitation industries. By analyzing the structural components of *Earthlings*, *Lucent* and *Dominion*, I have attempted to illustrate the multiple ethical problems that are at stake and to survey the possible effects of moral shock as a method that aims at pushing viewers into action. Representations of violence seeking embodied responses resort to shots and editing techniques that emphasize the quality and quantity of pain and suffering and that overwhelm the spectator with the spatial and temporal fractality of speciesism. This approach stands as one of the many rhetorical practices continuously explored in audiovisual discourses for animal advocacy, and its existence as part of a wider range of tactics attests to the movement's need for diverse, strategic forms of representation to connect with spectators with different interpretative processes.

In these types of films, the persuasiveness of moral shock relies on the trope of what is seen and what is concealed. In the introduction to this book, I explained that one of the characteristics of twenty-first-century animal advocacy documentaries is that they rested on the assumption that viewers had at least some previous knowledge about there being a public animal rights debate involving industrial practices. Jacques Derrida stated that

> [e]verybody knows what terrifying and intolerable pictures a realist painting could give to the . . . violence to which man has been submitting animal life for the past two centuries. Everybody knows what the production, breeding, transport and slaughter of these animals has become.
>
> ([2006] 2008, 26)

The films discussed in this chapter act upon viewers' liability to know (and to learn to know) more by assimilating other forms of sensorial kinship and viscerally dealing with the inapprehensible excess of victims. They deliver an abundance of violent images with the object of being spread and beheld, but with the ultimate purpose of extinguishing their representability of the present. These images should become an archive of the past, and as such they must follow a self-annihilating impulse – the type that institutionalizes memories of how things once were and should never be again. The residues of violent images must translate into action, and in that emotional precipice one perhaps confronts the most haunting thought of all: that we may be equipped with the mechanisms to live with ourselves and to continue as we were despite such knowledge and feelings of kinship.

Notes

1 Meanwhile, the participatory mode "emphasizes the interaction between filmmaker and subject" (Nichols 2001, 34). The poetic and reflexive modes engage with form with a higher aesthetic awareness of its apparatus as a mediator and fabricator of truths, with the former "emphasiz[ing] visual associations, tonal or rhythmic qualities, descriptive passages, and formal associations" (33) and the latter "call[ing] attention to the assumptions and conventions that govern documentary filmmaking" (34).

2 At a time of growing urbanization and the forging of Victorian sensibilities, the consolidation of experimental physiology as a prominent field of research contributed to the development of discursive explorations of the modern concept of cruelty (Turner 1980). Throughout the nineteenth century in Anglo-American culture, there was a flourishing of humane and anti-vivisection societies in response to the literature and to experiments and demonstrations carried out by vivisectors with live nonhuman animals. As a result, the image of the surgeon as a callous, merciless man of science blinded by the greed of knowledge became part of popular culture – a topic which has attracted feminist critique (Kean 1995; Lansbury 1985).

3 In this regard, *Unnecessary Fuss* must be historicized as part of a wider context of activism and tactics that involved PETA's formation of a coalition with other animal rights organizations, garnering as much media attention as possible, continuing to put pressure on Congress, and promoting civil disobedience (with a four-day occupation by activists of the National Institute of Health offices that funded Thomas Gennarelli's laboratory). The NIH had also had previous experience with animal rights activists' campaigns in the mid-1970s when the Museum of Natural History was targeted by the activist Henry Spira for their experiments on cats' sexuality, the research for which was regarded as useless by New York Congressman Ed Koch. Spira, today considered one of the founding members of animal rights activism, succeeded in stopping the NIH's funding of the museum, and would continue practicing strategies based on localized targeting and goal feasibility (such as against the Metcalf-Hatch Act, which allowed the state of New York to use unclaimed animals at shelters for research, or against the multinational company Revlon and their product testing protocols).

4 Ag-gag (agricultural gag) laws began to emerge in the United States in the 1990s to combat the rise of investigative research into farm practices without the consent of the owner, and similar initiatives have since then appeared in other countries such as Canada, Australia and France. While supporters argue that they safeguard farmers against activist misrepresentations of how they run their businesses, opponents regard them as a means to secure the impunity with which nonhuman animals are cruelly treated, as they "limit the ability of whistle blowers to inform regulators and the public of substandard production practices" (Riley 2022, 13).

References

Aaltola, Elisa. 2014. "Animal Suffering: Representations and the Act of Looking." *Anthozoös* 27 (1): 19–31. https://doi.org/10.2752/175303714X13837396326297.

Adams, Carol J. 1990. *The Sexual Politics of Meat. A Feminist-Vegetarian Critical Theory*. Continuum.

Adams, Carol J. 2003. *The Pornography of Meat*. Continuum.

Aranceta-Reboredo, Olatz, and Núria Almiron. 2023. "On Compassion, Animal Suffering, and Persuasion." In *Animal Suffering and Public Relations*, edited by Núria Almiron, 34–49. Routledge.

Arcari, Paula. 2022. "(Animal) Oppression: Responding to Questions of Efficacy and (Il) legitimacy in Animal Advocacy with a New Collective Action/Master Frame." *Animal Studies Journal* 11 (2): 69–108. https://doi.org/10.14453/asj/v11i2.3.

Barthes, Roland. 1977. *Image Music Text*. Translated by Stephen Heath. Fontana Press.

Berhmann, Marlene, and Galia Avidan. 2022. "Face Perception: Computational Insights from Phylogeny." *Trends in Cognitive Sciences* 26 (4): 350–63. https://doi.org/10.1016/j.tics.2022.01.006.

Bernatchez, Annie. 2024. "Emotional Reflexivity in the Animal Justice Politics of Sight: Embodied Moral Shock and Limit of the Emotional Repertoire." *Emotions and Society* 6 (1): 60–75. https://doi.org/10.1332/263169021X16740841736427.

Blewitt, John. 2010. *Media, Ecology and Conservation: Using the Media to Protect the World's Wildlife and Ecosystems*. Green Books.

Brottman, Mikita. 2004. "Mondo Horror: Carnivalizing the Taboo." In *The Horror Film*, edited by Stephen Prince, 167–88. Rutgers University Press.

Burgos, Javier, dir. 1986. *Hidden Crimes*. SUPRESS.

Burt, Jonathan. 2002. *Animals in Film*. Reaktion Books.

Burt, Jonathan. 2005. "Review of *A Day in the Life of a Massachusetts Slaughterhouse* (PETA Videos/1998)." *Society and Animals* 13 (4): 347–51.

Campion, Jane, dir. *The Piano*. Jan Chapman Productions.

Chang, Chia-Ju. 2016. "Wasted Humans and Garbage Animals: Deadly Transcorporeality and Documentary Activism." In *Ecodocumentaries. Critical Essays*, edited by Rayson K. Alex and S. Susan Deborah, 95–114. Palgrave Macmillan.

Delforce, Chris, dir. 2014. *Lucent*. Aussie Farms.

Delforce, Chris, dir. 2018. *Dominion*. Aussie Farms.

Derrida, Jacques. (2006) 2008. *The Animal that Therefore I Am*. Edited by Marie-Louise Mallet. Translated by David Wills. Fordham University Press.

Dunayer, Joan. 1995. "Sexist Words, Speciesist Roots." In *Animals and Women: Feminist Theoretical Explorations*, edited by Carol Adams and Josephine Donovan, 11–31. Duke University Press.

Eco, Umberto. 2009. *The Infinity of Lists*. Translated by Alastair McEwan. Rizzoli.

Fernández, Laura. 2020. "The Emotional Politics of Images: Moral Shock, Explicit Violence and Strategic Visual Communication in the Animal Liberation Movement." *Journal for Critical Animal Studies* 17 (4): 53–80.

Fernández, Laura. 2021. "Images that Liberate: Moral Shock and Strategic Visual Communication in Animal Liberation Activism." *Journal of Communication Inquiry* 45 (2): 138–58. https://doi.org/10.1177/0196859920932881.

Finn, Stephen Marcus. 2023. *Farmed Animals on Film. A Manifesto for a New Ethic*. Palgrave Macmillan.

Finsen, Lawrence, and Susan Finsen. 1994. *The Animal Rights Movement in America. From Compassion to Respect*. Twayne.

Fittock, Chris. 2018. "Could Shock Tactics Do More Harm than Good to the Vegan Cause?" *Open Democracy*, July 29. https://www.opendemocracy.net/en/transformation/could-shock-tactics-do-more-harm-than-good-to-vegan-cause.

Freeman, Carrie Packwood, and Scott Tulloch. 2013. "Was Blind But Now I See: Animal Liberation Documentaries' Deconstruction of Barriers to Witnessing Injustice." In *Screening Nature. Cinema Beyond the Human*, edited by Anat Pick and Guinevere Narraway, 110–26. Berghahn.

Goodall, Mark. 2006. "Shockumentary Evidence: The Perverse Politics of the Mondo Film." In *Remapping World Cinema. Identity, Culture and Politics in Film*, edited by Stephanie Dennison and Song Hwee Lim, 118–26. Wallflower Press.

Hansson, Niklas, and Kerstin Jacobsson. 2014. "Learning to Be Affected: Subjectivity, Sense and Sensibility in Animal Rights Activism." *Society and Animals* 22 (3): 262–88. https://doi.org/10.1163/15685306-12341327.

Jacopetti, Gualtiero, Paolo Cavara, and Franco Prosperi, dirs. 1962. *Mondo Cane*. Cineriz.

Jakobsen, Krisztina V., Cassidy White, and Elizabeth A. Simpson. 2021. "General and Own-species Attentional Face Biases." *Attention, Perception, & Psychophysics* 83 (1): 187–98. https://doi.org/10.3758/s13414-020-02132-w.

Jasper, James M. 2018. *The Emotions of Protest*. University of Chicago Press.

Jasper, James M., and Jane D. Poulsen. 1995. "Recruiting Strangers and Friends: Moral Shocks and Social Networks in Animal Rights and Anti-Nuclear Protests." *Social Problems* 42 (4): 493–512. https://doi.org/10.2307/3097043.

Jones, Huw D. 2022. "Achieving an Intimacy of Knowledge and Effect? The Impact of Documentary Films in Europe." *Journal of Audience and Reception Studies* 19 (1): 177–98.

Kääpä, Pietari. 2013. "Understanding the Audiences of Ecocinema." *Interactions: Studies of Communication & Culture* 4 (2): 107–11. https://doi.org/10.1386/iscc.4.2.107_2.

Kean, Hilda. 1995. "The 'Smooth Cool Men of Science': The Feminist and Socialist Response to Vivisection." *History Workshop Journal* 40 (1): 16–38. https://doi.org/10.1093/HWJ/40.1.16.

Lansbury, Coral. 1985. "Gynaecology, Pornography, and the Antivivisection Movement." *Victorian Studies* 28 (3): 413–37. https://www.jstor.org/stable/3827303.

Lowe, Brian M. 2006. *Emerging Moral Vocabularies. The Creation and Establishment of New Forms of Moral and Ethical Meanings*. Lexington Books.

McArthur, Jo-Anne. 2023. "'If People Want to Turn Away from the Subject Matter, You Have to Give Them a Reason to Turn Back to It.' Interview with Jo-Anne McArthur, Leading Animal Photojournalist." Interview by Claudia Alonso Recarte. *Journal of Animal Law and Interdisciplinary Animal Welfare Studies* 11 (RI §426985): 139–52.

Meat Institute. 2012–2018. "Glass Walls Videos." *YouTube Playlist*. https://www.youtube.com/playlist?list=PLkBbso1kwZ3bZTqN5MBLqHWGpRqPCH7gK.

Middleton, Jason. 2015. "Documentary Horror: The Transmodal Power of Indexical Violence." *Journal of Visual Culture* 14 (3): 285–92. https://doi.org/10.1177/1470412915607913.

Mika, Marie. 2006. "Framing the Issue: Religion, Secular Ethics and the Case of Animal Rights Mobilization." *Social Forces* 85 (2): 915–41. https://doi.org/10.1353/sof.2007.0017.

Monsó, Susana. 2021. *La zarigüella de Schrödinger.* Plaza y Valdés.

Monson, Shaun, dir. 2005. *Earthlings*. Nation Earth.

Monson, Shaun. 2019. "El director del documental *Earthlings* habla sobre su inspiración." *YouTube* video, 2:46. Posted by *Generación Vegana*, November 21. https://www.youtube.com/watch?v=w4-CWr3QkC4.

Morrison, Adrian R. 2009. *An Odyssey with Animals. A Veterinarian's Reflections on the Animal Rights & Welfare Debate*. Oxford University Press.

Morse, Tal. 2021. "Now Trending: #Massacre: On the Ethical Challenges of Spreading Spectacular Terrorism on New Media." In *The Age of Spectacular Death*, edited by Michael Hviid Jacobsen, 126–43. Routledge.

Natusch, Barry, and Beryl Hawkins. 2014. "Mapping Nichols' Modes in Documentary Film: *Ai Weiwei: Never Sorry* and *Helvetica*." *IAFOR Journal of Media, Communication and Film* 2 (1): 103–27. https://doi.org/10.22492/ijmcf.2.1.07.

Newkirk, Ingrid, and Alex Pacheco, prods. 1984. *Unnecessary Fuss*. PETA.

Nichols, Bill. 1993. "'Getting to Know You . . .': Knowledge, Power, and the Body." In *Theorizing Documentary*, edited by Michael Renov, 174–92. Routledge.

Nichols, Bill. 2001. *Introduction to Documentary*. Indiana University Press.

Nixon, Rob. 2011. *Slow Violence and the Environmentalism of the Poor*. Harvard University Press.

O'Brien, Sarah Jane. 2012. "Unnerving Images: Cinematic Representations of Animal Slaughter and the Ethics of Shock." PhD diss., University of Toronto.

O'Brien, Sarah Jane. 2023. *Bits and Pieces: Screening Animal Life and Death*. University of Michigan Press.

Parkinson, Claire. 2018. "Animal Bodies and Embodied Visuality." *Antennae: The Journal of Nature in Visual Culture* 46: 51–64.

Patterson, Charles. 2002. *Eternal Treblinka. Our Treatment of Animals and the Holocaust*. Lantern Books.

People for the Ethical Treatment of Animals. 2009. *Glass Walls* video, 12:45. https://www.peta.org/videos/glass-walls-2.

Pick, Anat. 2013. "Three Worlds: Dwelling and Worldhood on Screen." In *Screening Nature. Cinema Beyond Human*, edited by Anat Pick and Guinevere Narraway, 21–36. Berghahn.

Plantinga, Carl. 1996. "Moving Pictures and the Rhetoric of Nonfiction: Two Approaches." In *Post-Theory: Reconstructing Film Studies*, edited by David Bordwell and Noël Carroll, 307–24. University of Wisconsin Press.
Plantinga, Carl. 2009. *Moving Viewers: American Film and the Spectator's Experience*. University of California Press.
Prince, Stephen. 1993. "The Discourse of Pictures: Iconicity and Film Studies." *Film Quarterly* 47 (1): 16–28. https://doi.org/10.2307/1213106.
Rich, B. Ruby. 2006. "Documentary Disciplines: An Introduction." *Cinema Journal* 46 (1): 108–15. https://dx.doi.org/10.1353/cj.2007.0004.
Riley, Sophie. 2022. *The Commodification of Farm Animals*. Springer.
Schonfeld, Victor, and Myriam Alaux, dirs. 1981. *The Animals Film*. Slick Pics International.
Scudder, Joseph, and Carol Bishop Mils. 2009. "The Credibility of Shock Advocacy: Animal Rights Attack Messages." *Public Relations Review* 35 (2): 162–64. https://doi.org/10.1016/j.pubrev.2008.09.007.
Singer, Peter. (1975) 1990. *Animal Liberation*. Avon Books.
Smaill, Belinda. 2014–15. "New Food Documentary: Animals, Identification, and the Citizen Consumer." *Film Criticism* 39 (2): 79–102. https://www.jstor.org/stable/24777948.
Smaill, Belinda. 2016. *Regarding Life. Animals and the Documentary Moving Image*. State University of New York Press.
Sobchack, Vivian. 2004. *Carnal Thoughts: Embodiment and Moving Image Culture*. University of California Press.
Sontag, Susan. 2004. *Regarding the Pain of Others*. Penguin.
Spiegel, Marjorie. 1988. *The Dreaded Comparison. Human and Animal Slavery*. Heretic Books.
Stein, Jenny, dir. 2000. *The Witness*. Tribe of Heart.
Stokes, Ashli Q., and Wendy Atkins-Sayre. 2018. "PETA, Rhetorical Fracture, and the Power of Digital Activism." *Public Relations Inquiry* 7 (2): 149–70. https://doi.org/10.1177/2046147X1877021.
Terry, Mark. 2020. *The Geo-Doc: Geomedia, Documentary Film, and Social Change*. Palgrave Macmillan.
Turner, James. 1980. *Reckoning with the Beast. Animals, Pain and Humanity in the Victorian Mind*. Johns Hopkins University Press.
Wrenn, Corey Lee. 2013. "Resonance of Moral Shocks in Abolitionist Animal Rights Advocacy: Overcoming Contextual Constraints." *Society and Animals* 21 (4): 379–94. https://doi.org/10.1163/15685306-12341271.
Wrenn, Corey Lee. 2023. "Shocked or Satiated? Managing Moral Shocks Beyond the Recruitment Stage." *Emotions: History, Culture, Society* 7 (2): 298–321. https://doi.org/10.1163/2208522x-bja10045.

2 The "Biograph-able" Subject

Primates, Cetaceans and Anti-Captivity Rhetoric

One of the ways that animal advocacy documentaries denounce exploitation is by focusing on the story of an individual subject and what their life was like because of their contact with humans. When it comes to wild nonhuman animals, what the subject becomes in a lifetime of captivity is tantamount to what they cease to be and what they could have been. This chapter examines two well-known biographically oriented films, James Marsh's *Project Nim* (2011b) and Gabriela Cowperthwaite's *Blackfish* (2013), as examples of how narratives about the real lives of nonhuman animals engender questions about the ethics of captivity. It discusses what the elements of "biograph-ability" are when considering nonhuman animal subjects, and how these are connected to characterization and historicization. This biograph-ability navigates discourses and means of representation that cannot completely shed anthropocentrism and anthropomorphism. At the same time, through an ethical lens, viewers are compelled to recognize the right of nonhuman animals to autonomy and self-determination; in other words, their right to become their own "biographers." The films depict narratives of traumatized nonhuman protagonists that become martyrs of their species and, to do so, they extensively rely on the collective feeling of remorse that is confessionally shared by the humans that were directly or indirectly involved in their lives at one point or another. These human agents carry the parable and enable narrative continuity and historicization. But the films also offer representations that reassert nonhuman subjectivity in ways that inspire what Martha Nussbaum (2022) calls "wonder," and that ask that we remain curious about the minds of nonhuman animals not because of our marvel at interspecies contact and voyeurism (into which primates and cetaceans are forced through captivity), but because of how fascinating their consciousness actually is regardless of us humans.

One Among the Many: Delineating Individuality

In Chapter 1, I discussed how, in Shaun Monson's and Chris Delforce's encyclopedic films of the suffering of nonhuman animals, quantity (of suffering bodies and of sequential images and segments) was persuasively used to precipitate moral shock. I also mentioned that this principle of quantity could be subverted in two ways: on the one hand, repeated similarity of images can numb viewer response, causing

DOI: 10.4324/9781032661445-3

indifference and therefore de-authenticating the nonhuman animal experience. This counterproductive effect may bring viewers closer to an aesthetic response, following Elisa Aaltola's (2014) and Susan Sontag's (2004) theories. On the other hand, the inertia of sequencing can be interrupted by a single image that stands out for the viewer, who then perceives it as more compelling than previous or subsequent ones. Here the principle of quantity is mechanically interrupted, but the goal (ethical response) remains the same. For the most part, these types of shots are focused on a single individual, whose onscreen presence demands a proper historicization of their own unique subjectivity, regardless of any representativeness of their species.

This chapter focuses on the representation of individuality in films that can offer, in a way, an inverse approach to the principle of quantity: by reconstructing the biography of specific individuals and historicizing them in the wider context of human–nonhuman animal relations, these films become situated in a time and place that accommodate the viewer's sense of compassion differently. In contrast to the excess of bodies and systematized, graphic violence, these films circumscribe the story by foregrounding viewer identification with one nonhuman protagonist whose life experiences pace the unfolding of the plot. Implicitly, the narrative acts as a synecdoche connecting the individual to the whole (which are groups defined by their species or by their condition as victims of captivity), drawing on "typicality," that is, "the generalisability of the particular" (Corner 1996, 49). This allows the viewer to balance the singularity of the protagonist's experience with the collective vulnerability of the group, and brings an ethical perspective to the fixation of the animal-industrial complex on biologization, that is, the definition, conceptualization and valuing of nonhuman animals on the basis of how they emblematize the biological, genetic and behavioral traits ascribed to their species. Biologization essentializes nonhuman animals; it establishes an identity for them based exclusively on the characteristics and expectations associated with their biology, objectifying them (Noske 1997; Colling et al. 2014, 66–67). The animal-industrial complex capitalizes on these traits in numerous ways, enhancing them genetically and in their lobbyist discourses (Aranceta-Reboredo et al. 2025). The films discussed here subvert such biologization by deploying an ethical lens and combining scientific evidence of species subjectivity with the individuality of the protagonists.

The biographical structure also offers a tangible narrative temporality precisely because life stories operate upon the continuity of a historical chronology that tends toward familiar plot developments based on a conflict with a rising action, climax and resolution. Life stories are dramatized into an action resembling that of Freytag's Pyramid. Meanwhile, in compilatory, "omnibus documentaries" (Finn 2023, 75–88) of the type discussed in Chapter 1, there is no narrative conflict in the traditional sense of the term: the graphic display of images of violence is organized according to certain criteria (types of industries or species, for example) and the linearity of action is conspicuously flat, in the sense that all images, because of the similarity of their composition and how they are sequenced, remain within the spectrum of high tension. These films sacrifice traditional narrative conflict in favor of a reliance on uniform, compelling deixis that stimulates the somatic response to images of physical pain and psychological suffering. Their temporal and spatial ambition, furthermore, disperses human accountability in the animal-industrial complex, where everyone involved in the chain of production and consumption has their share of responsibility. Documentaries that are biographically oriented, on the other hand, place their characters in a specific time–space axis, and their lives are chronologized into a narrative arc rising above the temporality cadenced by the machinery of animal agriculture and the meta-structure of compilatory sequencing. This situatedness allows a clearer delimitation and contouring of the antagonist, and it is important to remember that the synecdochical powers of the nonhuman animal protagonist also apply to the antagonist: the identifiable faces, businesses and corporations that in these films represent the exploiters and oppressors, stand as the visible side of a complex perpetuated thanks to the participation of millions of off-screen consumers. Biographically oriented films, in other words, are tacitly aware of the universality of their characters, action and tropes.

This chapter looks at the concept of biograph-ability, asking why and how the lives of selected individual nonhuman animals become particularly compelling and make for the proper substance of advocacy narratives. It examines the devices employed to represent such lives, and how the principles of animal ethics are developed in connection to their structures. So-called animal autobiographies have, since the nineteenth century – and thanks to classics such as Anna Sewell's *Black Beauty* (1877) – held a popular spot in (usually) children's and young adult literature, which is already revealing of how attempts at representing nonhuman animal thoughts and subjectivity have traditionally been regarded as infantile. Animal advocacy documentaries that are biographically oriented have inherited some aspects of these representations, including the type of dramatic action mentioned earlier to map and pace life stories, but they have also rejected some of its most salient features – most notably, the nonhuman animal narrator, an anthropomorphic gesture that is too obvious an assault on the genre's claims to truth.

A clarification with regard to the terminology is in order: I am inclined to speak of "biographically oriented documentaries" in the case of nonhuman animal subjects as opposed to "biographical documentaries" to emphasize the very limitations of such a subgenre when focusing on a nonhuman animal protagonist. In animal advocacy documentaries, I argue that what qualifies as biographically oriented films are those that represent the life of an individual nonhuman animal as affected by their intersection with and immersion in the human world. Evidently, whether human or nonhuman, the subject of a biographical enterprise is always liable to distortion, exaggeration or misrepresentation when placed at the heart of a film, because the very impulse of documentary, as described in the introduction, does not just preclude objectivity, but in fact questions its very existence, offering versions of "truths" as composite interpretations instead. The camera "produces an indexical record of what falls within its visual field. As an anthropomorphic extension of the human sensorium the camera reveals not only the world but its operator's preoccupations, subjectivity and values" (Nichols 1991, 79). This serves as a reminder that we can only look at *representations* of the life of a subject, and when faced with the species divide, these representations are intrinsically problematized by the limits of our knowledge, not so much of the context in which such life unfolded, but of the subjective and phenomenological experience (Nagel 1974). Biographies, therefore, "remain external to the cognitive experience of the animals' worlds and their relations to it" (Krebber and Roscher 2018, 2). The species divide calls for an extension of Jane Chapman's question as to whether filmmakers can represent the cultural other without succumbing to "stereotyping" and "a type of 'visual imperialism'" (2009, 34). Jennifer Ladino in fact refers to the "speciesist camera" that "sees animals through human eyes, distorting their behaviors with a social lens that is fraught with both ideological and generic expectations" (2013, 130). To call the documentaries at hand "biographically oriented" as opposed to "biographical," therefore, acknowledges the forces of anthropocentrism that cannot be detached from the camera.

Project Nim and *Blackfish* were not actively marketed as biopics, although the term (as well as "biography") did occasionally appear in promotional blurbs and short summaries and in film festival webpages, especially in the case of the former. Marsh based his film on Elizabeth Hess's *Nim Chimpsky: The Chimp Who Would Be Human* (2008). As he claimed in an interview for *Empire Magazine*,

> I read the book and I was just gripped by the surprising twists and turns of [Nim's] life story. And the life story is also very good at genre: it has a beginning, middle and end. You're given that in a life story – it's a biopic. So I thought it would be interesting whether you could actually . . . use a film biography to tell the story of an animal's life.
>
> (Marsh 2011a)

Cowperthwaite's inspiration for *Blackfish*, meanwhile, was an *Outside Magazine* article by Tim Zimmermann titled "The Killer in the Pool" (2010). *Project Nim* and *Blackfish* use biography as both structure and pretext. The films respectively engage with a particular event in recent history: the launching, development and aftermath of Dr. Herbert Terrace's scientific experiment to teach sign language to a chimpanzee (beginning in the 1970s), and the tragic death of SeaWorld Orlando senior trainer Dawn Brancheau in February 2010. The events are strategically contextualized within the narratives of a life of hardships and trauma suffered by both Nim and the orca who took Brancheau's life, Tilikum. Through a "string-of-interviews" (Nichols 1983) that effectively replaces the need for a voice-over, the lives of Nim and Tilikum are stitched into patterns of trials and tribulations, as their subjectivities are interpreted by these "authorities" and eloquently supported by archival footage, photographic material and – in the case of *Project Nim* – re-enactments.[1]

It must also be said that biographical reconstructions of nonhuman individuals need not, of course, always lead to these films' stylistic, cinematographic or editing choices. For example, a comparison between the aesthetics of Victor Kossakovsky's *Gunda* (2020) or Andrea Arnold's *Cow* (2021), which can arguably also be characterized as biographically oriented films, and *Project Nim* and *Blackfish*, immediately reveals fundamental creative differences on the part of the filmmakers. While the cinematography of the former two is less conventional and more reflexive, reducing the onscreen human to either symbolic indexes or fleeting presences, the latter two essentially construct their narratives around human testimonies. Because Chapter 4 already discusses *Gunda* and *Cow* from a gender studies perspective, I shall not analyze them here. Nonetheless, it is important to keep in mind that although the biograph-ability strategies used in *Project Nim* and *Blackfish* are indeed effective techniques, they are not the only ones. Other documentaries explore in their own ways the representation of their protagonists' subjectivity, and often point to the filmmakers' conscious attempt to distance themselves from anthropomorphism.

Aside from the question of *whether* a nonhuman individual is truly biograph-able, there is also the matter of *which* nonhumans awaken our interest in their lives. Film can, of course, potentially motivate viewers to become curious about different types of lives (for instance, wildlife documentaries, as discussed in the introduction, offer entertaining and palatable reconstructions, often attuned to a biologization of their subjects). But while technological advances have made it possible to experiment with the representability of the life experience of a whole range of species whose size, accessibility or environment, until very recently, impeded satisfying visibility (think of insects, deep-ocean biodiversity or even Computer-Generated Imagery [CGI] dinosaurs) (Richards 2010; Scott 2010; Louson 2021), the cultural appreciation of some species can be particularly impacting and contribute to a predisposition to know more about them. This is the case with charismatic fauna. These types of wild nonhuman animals "[seize] the public imagination or sympathy for one reason or another" (Beatson 2011, 38), and they are often referred to as "megafauna" to indicate the importance of size (and hence of an appreciable visibility), which boosts people's interest. They have "compelling features and interpretations to make them newsworthy" (Maynard 2018, 186–87). Charisma is a decisive element in conservation and the promotion of flagship species that attract people enough to care (Freitas Torres et al. 2018, 436; Howlett et al. 2023, 634) does elicit reconsiderations of how to balance public perceptions of such species with regard to other endangered (Colléony et al. 2017) or noncharismatic ones (Hosey et al. 2020). Basing their research on their reading of Jamie Lorimer's (2007) study in the context of biodiversity conservation in the UK, Frédéric Ducarme et al. (2013) summarize the factors that are believed to heighten charisma. They refer to "'*detectability and distinctiveness*' (i.e. fame), '*socio-economic biases*' (the way societies see each animal and its reputation), aesthetics (on a cuddly charisma – feral charisma *continuum*), and '*potential to generate satisfaction*' (which represents its interest for scientists, intellectuals and curious people)," but acknowledge that charisma is ultimately "very relative" and "able to be enhanced or constructed artificially" (2013, 2; emphasis in original). Charismatic (mega)fauna spark interest by meeting certain criteria at the right time and place, and their popularity is affected by trends and cultural whims that can have long-lasting effects and can reinvent a species' reputation.

In Western popular culture, some cetacean and primate species are generally regarded as part of the charismatic (mega)fauna lore. With growing scientific evidence of their intelligence and their rich emotional and social lives, the fascination they excite is both reflected in and nurtured by subsequent anthropomorphic representations that seek to satisfy a human desire for interspecies connections. It is this longing for connection (artificially stimulated by mainstream visual discourses such as films, television shows, advertisements, tourist attractions, or zoos and aquaria, all of which contribute to the aforementioned factors heightening charisma) that keeps them newsworthy and popular. And it is important to remember that charisma and popularity are mostly sustained because human interpretations (whether these be more scientific, academic, ideological or sentimental) offer "readings" of behaviors that spark further interest in interspecies connections and contact. As Julia Corbett ascertains, when it comes to newsworthiness, wildlife becomes of interest "either when someone is making claims about them, or when the boundaries – be they symbolic or real – between humans and untamed animals are somehow breached." She goes on to add that

> [e]ven purely human interest stories about wildlife involve a contact with humans and an overlapping of the domesticated and undomesticated realms. Animals doing what animals do normally quite separate from humans are topics not well-suited for the average newspaper or TV newscast.
>
> (Corbett 1995, 399)

Much of Nim and Tilikum's biograph-ability is grounded in their status as charismatic megafauna. Graham Huggan notes that, in the case of Cowperthwaite's film, "emotional appeal" is organized

> around the figure of the orca as a large charismatic animal whose intelligence and physical force are rendered proportionate with its suffering and whose immediate recognizability, together with its symbolic resonance, makes it a particularly attractive candidate for financial support.
>
> (2017, 297)

Meanwhile, the phylogenetic similarities between humans and primates (and with chimpanzees in particular given that they are, alongside bonobos, the closest related species to humans) make them regular recipients of collective and public affection. The popularity of the work of the late Dr. Jane Goodall and Dr. Frans de Waal, among other primatologists, was instrumental in shaping positive public perceptions of chimpanzees and great apes.

Another reason that made Nim and Tilikum particularly biograph-able is that they had a traceable history in the media that molded their public image and personas. As reflected in *Project Nim*, Terrace's experiment to discover whether chimpanzees had grammatical capabilities became, for a while, a hot topic in the mid-1970s, with a front cover on *New York* magazine and coverage by *The Daily News* and other media outlets and television programs. For a brief moment, Nim became somewhat of a celebrity, an actual flesh-and-bone chimpanzee that promised a communication breakthrough to a society enthralled by the sci-fi optics of the *Planet of the Apes* franchise and its merchandizing.[2] Tilikum, on the other hand, had become (in)famous with the death of Brancheau, pushing SeaWorld into utter crisis mode – a crisis that, as we will see, wildly escalated with the release of *Blackfish.* There was a precedent, in other words, of media accreditation of Nim and Tilikum as individuals, however distorted or anthropomorphized such portrayals were. As Jason Sperb puts it, "[w]ithout the sensationalistic elements of Brancheau's death already in play, would *Blackfish* have found an audience?" (2016, 217). Credentials of individuality in media records are garnered because the stories surrounding the subjects historicize them, bringing them more clearly to the forefront, while the mass of other individuals of their species, regardless of their own charisma, fade into a background compiled of general ideas and beliefs regarding the group at large. Media salience is made possible because there is a historicization of the singleness and never-before-ness of an event. Even if Tilikum had previously been involved in the deaths of two other people, it is the dramatic killing of a woman in the corporate giant that is SeaWorld that earns him the profiling and the notoriety that comes with it. In the same way, Terrace's experiment stood at the cusp of science and the nature-versus-nurture debate, making Nim the "face" of futuristic possibility.

Because the public tends to misconstrue nature as a space where wildlife is bound to marked, repetitive cycles, and where new generations that are hardwired by their antecessors' instincts confront the same challenges, proper historicization beyond the wide spectrum of evolution is not necessarily factored in. There is a sense of timelessness to the appreciation of wildlife, even in zoos and aquaria, where nonhuman animals are traded and replaced to create an aesthetic that pleases visitors who, in all likelihood, find it extremely hard to distinguish between one individual and another without proper assistance. The tragic and traumatic circumstances of Nim and Tilikum – as captive, nonconsenting nonhuman animals that are thrust into the mayhem of exploitation industries – are conducive to their historicization as individual, "special" figures. Such status contributes to their biograph-ability in Marsh's and Cowperthwaite's films. This meaning-making of historical material again echoes Erica Fudge's claim that "the history of animals is in reality the history of human attitudes toward animals" (2002, 6).

From a philosophical point of view in animal ethics, the completion of a life story is connected to self-awareness and consciousness. Biograph-ability, in this sense, also demands consideration, through an ethical lens, of the environmental circumstances in which the life of the subject unfolds. Captive and nonconsenting nonhuman animals (and potentially all victims of the animal-industrial complex) have their "biographies" interrupted by human activity. If prevented from thriving in their own lifetime, the subject is being harmed because they are being deprived of what matters to them. In a way, biography is connected to Bernard Rollin's notion of *telos*, by which he means

> [t]he set of needs and interests which are genetically based, and environmentally expressed, and which collectively constitute or define the "form of life"

> or way of living exhibited by that animal, and whose fulfilment or thwarting matter to the animal.
>
> (1998, 162)

According to this view, in order for nonhuman animals to be morally considered, their sense and experiencing of "mattering" require a certain level of consciousness and awareness: "[T]o enjoy moral status, an entity (i.e., an animal) must have the kind of *telos* whose violation creates some negative mode of awareness in the creature in question" (Rollin 2016, 53). Martha Nussbaum's Capabilities Approach also considers the form of life that is specific to both species and individuals, upon whom she places great emphasis as agents entitled to dignity. The Central Capabilities, which are conceived as "*substantial freedoms*" and "core entitlements" comparable to "a list of fundamental rights" (Nussbaum 2022, 80; emphasis in original), grant the subject the "opportunity to choose to act, in a specific area of life deemed valuable" (86). As such, these Capabilities acknowledge the different ways that sentient creatures choose to live their life; that is, there is an understanding of the heterogeneity of experience and ways to act upon the world. What is relevant to each creature is their own form of flourishing, and that should leave room for individual differentiation in a world where they can choose how to act. Nussbaum's Capabilities Approach, furthermore, demands that we accept the narrative structure of the life of sentient creatures that are aware of the passage of time:

> [L]ives unfold in time, and for a creature highly conscious of time and who lives, as humans typically do, in both the past and the future as well as the present, death can be a harm by interrupting the temporal flow.
>
> (163)

The Capabilities Approach "ask[s] us at all times to consider the creature's whole form of life, including its temporal shape" (163).

Thus, anthropocentric and anthropomorphic forms of storytelling affect and determine the biograph-ability of the nonhuman animal subject and their individualization, but from an animal ethics perspective, biograph-ability is connected to their moral status as conscious subjects whose lives unfold in time and in a world upon which they choose to act. These positions are difficult to reconcile; while one biographical impulse seeks to ascertain anthropocentrism, the acknowledgment of biography through the lens of animal ethics foregrounds the individual and the species of which they are part regardless of human interests, thus operating on the premise that each nonhuman animal is not a means, but an end unto themselves. Shortly, I shall discuss *Project Nim* and *Blackfish* as biographically oriented films constantly negotiating these tensions: on the one hand, the lives of Nim and Tilikum are often reconstructed in a way that prioritizes the penitence of the human interviewees, and in their confessional state, they acquire the authority to infer the subjective experience of the nonhuman animal subjects, but also, paradoxically, to acknowledge the limits of their inference. On the other hand, the anti-captivity message insists on the possibility of a narrative structure where the protagonists would be the sole agents: a state in which they, as individuals able to choose and act upon the world, would be their own "biographers."

The Interior Landscape of Trauma

Project Nim and *Blackfish* have a number of things in common as biographically oriented documentaries: (1) they both revolve around the life of an individual whose species qualifies as charismatic megafauna; (2) they situate a historical event within the logical course of a life story in captivity; (3) structurally speaking, their life stories are traced back to a traumatic beginning when the subject is introduced and subjugated to a human world; (4) they are informed by a "string of interviews," rather than a voice-over, with a cast of authority figures mostly made up of people who were involved in the protagonists' lives; (5) many of these interviewees are characterized by a discourse of remorse and regret, positioning the narratives as rituals of confession and penitence; (6) as cautionary tales, they historicize Nim and Tilikum as individuals representing real-life tragic examples of how *not* to treat intelligent, sentient beings; (7) they envision potential flourishing lives for their subjects thriving as individuals – lives where they can exercise their freedom to choose, to move and to socialize; and (8) Nim and Tilikum, individualized as they are, are also martyrs epitomizing the perils faced by their entire species. The message of advocacy is present in both films, even though neither Marsh nor Cowperthwaite identified as animal activists (Marsh 2011c; Cowperthwaite 2014) and their previous work in the documentary genre had little to do with the question of nonhuman otherness.[3]

Project Nim enjoyed vast critical success and was nominated for various prestigious awards (BAFTA, British Independent Film Awards, and Critics Choice Awards, among others) and won the 2012 Directors Guild of America award in the category of Outstanding Directorial Achievement in Documentary. The film follows the life of Nim Chimpsky as the subject of Terrace's experiment. Born in November 1973 at the Institute for Primate Studies (IPS) in Oklahoma, Nim is separated from his mother at two weeks old and sent to live with a former student of Terrace's, Stephanie LaFarge, and her large family. The LaFarge household where Nim spends his earliest developmental stage is characterized as an accommodated hippie-like environment of the 1970s. Unsatisfied with the lack of methodology and scientific rigor, Terrace eventually assigns Nim to a new team of caretakers and teachers of sign language – Laura-Ann Petitto (characterized as a very young and ambitious student), Bill Tynan, Joyce Butler and Renne Falitz – who settle at Delafield Estate, a property in Riverdale owned by Columbia University. For two years Nim seems to thrive at Delafield, quickly and steadily learning dozens of signs and communicating his most immediate demands to his handlers. However, Nim is getting bigger and increasingly difficult to handle, to the point where his violent episodes (banging Petitto's head on the pavement and biting Falitz's face) convince Terrace to finish the data-collecting stage of the experiment and return Nim to the IPS, directed by Dr. Lemmon, who is characterized as a cold-hearted scientist and remains largely unpleasant for both viewers and interviewees.

Nim goes from a life of "privileged captivity," raised like a human baby, to living in a small cage, being handled with a cattle prod and wearing a collar with a lock around his neck. The inability to handle chimpanzees once they reach maturity is an all-too-common result for anyone attempting "to use human-rearing tactics to control the infant," as "the seeds of developmental disorder are sown" (Ross 2014, 70). Nim's lack of contact with other chimpanzees during his youth destroys his socializing skills with members of his species, and the film suggests that the only reason he made it through this traumatic change was the friendship he developed with Bill Ingersoll, a research student at IPS. Ingersoll takes Nim on walks, shares pot with him and together they form a companionship. In answer to Terrace's claim that the scientific project had failed (or, more accurately, that it seemed to prove that chimpanzees did *not* have grammatical capacity but were instead responsive to forms of learning similar to conditioning),[4] Ingersoll states in the film:

> I didn't care about the language . . . after a while it didn't matter to me. He might not have had sentences or grammar, but there's no question that there was communication going on, and I saw it clearly.

In May 1982, New York University's Laboratory for Experimental Medicine and Surgery in Primates (LEMSIP) acquires a large colony of primates from a financially troubled Dr. Lemmon, and so Nim and other chimpanzees are shipped back to New York. As testified by LEMSIP Research Veterinarian Dr. James Mahoney, the conditions in the laboratory were inhumane, with chimps used as test subjects for vaccines and kept in small cages. In a smart legal move, attorney Henry Herrman, with the help of Ingersoll, manages to have Nim released from the laboratory, and animal advocate Cleveland Amory purchases him and takes him to Black Beauty Ranch in Texas, a sanctuary for equines. Unfortunately, despite the comparatively much larger cage built for him at the ranch, Nim remains depressed and isolated. His frustration climaxes when Stephanie LaFarge and her daughter visit him. Upon entering his cage despite his visibly agitated state, LaFarge is violently attacked by Nim. The film comes, in this way, full circle with the opening re-enactment of Nim's first abduction from his biological mother, and the successive abandonments he suffered as he was moved from place to place. The film, nonetheless, ends on a more positive note as the last five minutes condense how, a year after Nim was brought to the ranch, a female chimp was purchased to be with him. Ten years later and once more with the help of Ingersoll and the collaboration of Mahoney, several chimpanzees from LEMSIP are taken to the ranch – a vast social improvement that Nim will only enjoy for the last five years of his life, as he dies of a heart attack in May 2000, at the age of 26.

Project Nim employs unsubtle melodramatic music to heighten suspenseful and tragic scenes, often combined with the poignant sounds of chimpanzee screams (conveyed as indexes of panic or distress). The amount of archival material retrieved is outstanding: videos collected during the research of the project; countless photographs inserted in crucial narrative moments; press clips; Ingersoll's own visual memorabilia, and recordings from LEMSIP and Black Beauty Ranch. The film reconstructs the past with a skillful intercalation of these (moving) images. Effective rhythmical energies are ignited with the panning and zooming over still photographs (a low-key reminiscence of the "Ken Burns effect"), which are woven into the map of the interviewees' accounts. These are the compositional elements arranging the narrative of Nim's life.

Blackfish draws on similar techniques affecting action and mode, to the point where the film crosses the threshold between genres and blends together elements of melodrama, wildlife film and true crime, adding occasional animation and media clips. While the film does not make much use of still photographs, its consistent reliance on the moving image is particularly striking because of the contrast it marks with the theme of captivity, where movement is violently constrained and limited to artificial spaces. The restriction of captive orcas' movements is emphasized using marine park promotional videos and footage together with recordings of underwater views through tank glass. These images are antithetical to recordings of wild orca pods, which authenticate and reaffirm the *telos* of cetaceans as creatures whose wellbeing rests on spatial freedom. *Blackfish*'s captive-free antinomy is exacerbated in the film's promotional image and variety of DVD covers, which feature a frontal picture of Tilikum's face, disproportionately large relative to the frame. This motif of constriction conceives an alternative visual and anti-speciesist language to represent captivity that echoes Joan Dunayer's (2001) proposal to replace the term "aquarium" with "aquaprison."

Structure is extremely important in *Blackfish.* In *Project Nim*, it is the re-enactment of Nim's removal from his mother that sets the tragic tone of the film in the pre-title sequence. *Blackfish*'s prelude, on the other hand, offers not the origin of trauma, but a glimpse of its consequences: the calls on February 24, 2010, to Orange County Fire Rescue and the Sheriff's Office are played while short clips of a trainer underwater in an orca tank are shown. "A whale ate one of the trainers?" asks the receptionist in disbelief. "That's correct," the caller from SeaWorld replies, and the sequence reaches momentum with the image of an orca pushing the trainer from below into the air for an impressive dive for a show. Both pre-title sequences exemplify documentary film's stimulation of "epistephilia (a desire to know) in its audience" (Nichols, 2001, 40). *Blackfish*'s adherence to true-crime conventions attracts viewers to the circumstances of what is packaged as a mystery: is the tragedy the result of an accident or a crime? The information delivered by the calls is boxed in species discourse duality: a whale (the perpetrator) has eaten a trainer (the human victim). As the documentary unwraps its character portrayal of the individuals behind these categories, a story of trauma unfolds, and the viewer is provided with the pieces that map the effect back to the cause, and that suggest that Brancheau's death was not an accident, but the expected outcome after decades of Tilikum's mistreatment and abuse by the entertainment industry.

After the opening sequence, the film introduces its core set of interviewees. These are former SeaWorld trainers who knew Tilikum and Brancheau and declare to have been completely unprepared for the job, thus initiating the film's accusation that SeaWorld knowingly put their workers in danger by withholding vital information on the nonhuman inmates' prior histories. There is an all-American-girlness to Brancheau's persona that not only renders her likable to the viewer but also fuels the proclivity of true crime and sensationalist media toward young, attractive, white middle-class female victims (Rickard 2023, 6, 19). The film again reinstates the theme of her death with the Occupational Health and Safety Administration's (OSHA) investigation of SeaWorld and the risk that trainers are subjected to when put in the water with whales. The controversy leads to media frenzy over safety and the ethics of captivity. As journalist Jane Velez-Mitchell quips, "If you were in a bathtub for twenty-five years, don't you think you'd get a little irritated, aggravated, maybe a little psychotic?". From there on, testimonies of the people directly or indirectly involved recount Tilikum's miserable life in captivity. The film contests the propagandistic marine park narratives of happy nonhuman animal attractions by providing scientific evidence of orca cognition and sentience, voiced by cetacean advocate and neuroscientist Lori Marino and whale researchers Howard Garrett and Dave Duffus.

Although the film does not produce any witness of Tilikum's 1983 abduction in Icelandic waters at the age of two, the testimony delivered by a participant in a 1970 capture of orca calves in Puget Sound gives a vivid representation of what Tilikum's experience must have been like. This marks the beginning of a wretched, lifelong seclusion in small pools and tanks with a complete lack of stimulation (first at a shoddy aquarium in British Columbia called Sealand, and then at SeaWorld Orlando), where frustration caused by the lack of space and the cultural incompatibility between orcas leads to aggression among them. As perpetrator of the death of Brancheau (and, the viewer learns, of another trainer at Sealand and possibly of one mentally unstable man who broke into SeaWorld), Tilikum is also simultaneously portrayed as the victim. Among the inaccuracies purported by SeaWorld are the claims that orcas live longer in captivity, that the rate of collapsed dorsal fins is similar in the wild, and even that perhaps it was Brancheau's lack of caution that got her in trouble in the first place. All these elements add a piece to the mystery that is Tilikum's mind, and so "only after establishing the orca's troubled history and the risks of working with him does the film finally show the attack on Brancheau, creating a powerful emotional impact for the audience" (Ryan et al. 2023, 4).

Blackfish premiered at the 2013 Sundance Film Festival and was lauded by critics. The film was distributed by CNN and platforms, including Netflix, and its widespread popularity led to what became known as "the *Blackfish* effect": the downward spiral of SeaWorld's credibility for stakeholders and visitors, a fascinating case study of how corporations can persistently fail in their management of legitimacy crises due to unfortunate message content, poor timing for response and incompetent use of online networks. SeaWorld's previous refusal to participate in the film did not help either, as it only fueled suspicions about their accountability. To assuage shareholders and the public, they eventually turned to ethics-washing and decided to phase out orca breeding programs, terminate orca shows and focus on education and conservation initiatives that acknowledged the dispute over killer whale lifespans in captivity (Callaway 2016). The "*Blackfish* effect" on SeaWorld led to million-dollar losses in revenue, rapidly declining attendance from visitors, public shunning from celebrities and strategic cyberactivism. These have all been sufficiently documented in studies on the deliberate misuse of science for business purposes (White 2017) and on the actors' attempts to convincingly control the narrative of truth in issue management and crisis framing (Chattoo 2016; Woods 2018; Parsons and Rose 2018; Javanaud et al. 2018; Waller and Iluzada 2020; Massey and Randriamarohaja 2019; Boissat et al. 2021), so I shall not dwell on it here. Suffice to say that the tremendous impact of the film serves as a model for strategizing about future animal advocacy documentaries (Merskin and Freeman 2024), and that its success may have been symptomatic of the extent to which the public was already becoming increasingly uncomfortable with killer whale shows and captivity – a collective feeling that has continued to be the object of assessment in fairly recent studies (Wasserman et al. 2018; Naylor and Parsons 2019; Bruck 2024). Public perceptions of killer whales had indeed come a long way from their representation as vengeful monsters in the 1977 *Jaws*-inspired *Orca*, directed by Michael Anderson. The success of the 1993 family film *Free Willy*, directed by Simon Wincer, which exalted the companionship between a troubled preteen and an orca at a local aquarium, evinced the incongruency of using a captive whale to create a passionate anti-captivity narrative (Alonso-Recarte and Ramos-Gay 2022, 423). Scientists were also making significant discoveries on the emotional and social intelligence of these cetaceans (Rendell and Whitehead 2001; Strager 2023). But the impact of *Blackfish* also illustrates public perceptiveness and responsiveness to life stories of charismatic megafauna and suggests that narrative control of nonhuman animal subjectivity can trigger the type of emotions that can lead to change – the normative response (Aaltola 2014; Sontag 2004), or what Nussbaum (2022) calls "Transition-Anger."

As should be clear by now, the biographical concessions of *Blackfish* and *Project Nim* rely on the compositional logic of a single narrative pieced together by the testimonies of interviewees, most of whom build their discourse around remorse. Like the diver involved in the capture of orca calves in Puget Sound, who regrets having "kidnap[ped] a little kid away from her mother," Stephanie LaFarge similarly describes Nim's separation from his biological mother (and eventually from her) as a cruel act of robbery. In both films, it is the tracing of trauma back to an infant stage when the subjects were abruptly abducted from their biological mothers that lights the spark climaxing in their respective attacks on their surrogate human ones: Brancheau and LaFarge. Baby-stealing carries on as a motif in *Blackfish* as the trainers recall the grief of the orca mothers at SeaWorld whose calves were sold off. As reported by researcher Howard Garrett in the film, wild orca offspring remain with their mothers for the rest of their lives. Chimpanzees stay close to theirs for upwards of five years (Ross 2014, 58).

The orca trainers and Nim's teachers regret their negligence and silence over the years and express their repentance. Former director of Sealand Steve Huxter recalls how keeping the orcas in such small confinements "didn't feel good; it just didn't. And it was just wrong." Similarly, Joyce Butler laments that: "I strongly believe that we made a commitment to [Nim] and we failed. We did a huge disservice to that soul, and shame on us." Even James Mahoney somewhat cleanses his negative image as an animal experimenter by helping to relocate the LEMSIP chimpanzees. Despite his comparatively brief appearance as an interviewee, Mahoney becomes an unexpected commendable figure: "There's no way, in all honesty, there's no way you can carry out research on animals and for it to be humane. It can't be humane because you already put them in a cage" (Marsh 2011b), he says. Contrary to the more unsympathetic portrayal of Terrace, who also expresses some hints of regret at having returned Nim to Oklahoma, Mahoney becomes redeemable because, like Ingersoll, he acts to improve the well-being of the chimpanzees (even though we may assume that he remains active in his profession). His relevance as a fundamental agent in Nim's life is acknowledged with Marsh's decision to leave to him the final statement in the film: "Chimps are truly wonderful animals. They're very forgiving, the vast majority of them. They'll forgive you." Because of Mahoney's background as a scientist, his words more readily resist being perceived as anthropomorphic. More importantly, *Project Nim* ends on a hopeful, ritualistic note of forgiveness with its promise of relief and the restoration of one's conscience. Similarly, the end of *Blackfish* shows the group of former trainers beholding wild orcas from a boat, and delivers a scene of final purification from the wrongful deeds inflicted upon SeaWorld's cetacean inmates. The redeemability of the ex-trainers and Mahoney is one of the features that more clearly qualifies the films as advocacy narratives, given that they are figures that are "forgiven" because they have actually and meaningfully engaged in some form of action to right their wrongs. The sentimentality of the *Blackfish* scene suggests that the film treats penitent testimony as a redeeming form of action itself, while *Project Nim* is perhaps more demanding of its interviewees. At times affectless, at times aseptic and impassive, and at times playfully conscious of the curious group of human specimens at its disposal, the camera captures signs of authentic regret, whether big or small. But Marsh seems less interested in meriting these testimonies as actions (of the type undertaken by Mahoney and Ingersoll) than in treating them as objects of fascination. The personal romantic, sexual and familial relationships between the interviewees harvest a narrative of messy human affairs that foreground feelings of longing and desire. But their vulnerability and fallibility are not so much looked upon with forgiveness as they are sought to place an unsuspecting Nim at the center of a storm of clashing human interests, egos, ambitions and hubris. Catching on to the film's emphasis on his own share of carelessness and negligence, Terrace expressed disagreement with his portrayal as an "absentee landlord" that "didn't care emotionally about [Nim]" (2011, 173).

In her study of private trauma in true-crime documentary, Anita Biressi discusses how "trauma documentaries across the spectrum from the 'serious' to the 'popular' attempt to represent individual and also collective fantasies: mythological structures that explain culture to itself, structures which viewing subjects will recognize and with which they will connect" (2004, 405). The strings of interviews effectively represent Nim and Tilikum as at once individuals and archetypes of captive victims, but they also reflexively affect the cultural recognizability of the interviewees themselves. The impression is that at times the films seem to be more about the shared trauma of the people involved than the actual sentient subjects that remain at the center of abuse, even if the viewer's moral focus on the nonhuman protagonists is steadfast. As Cowperthwaite admits in an interview:

> I did tell [the story] through the former trainers' perspective. I saw them as my apostles . . . And I just thought they are incredibly very credible. Been there, worked there, believed the hype, and had become disillusioned. And so, they were my storytellers. They are my film, essentially.
>
> (2014)

Cage and Wonder

The way the films arrange the raw material – the interviews – into the apparatus of biography prompts epistemological questions as to the true knowability of the nonhuman subject. Perhaps because of this reliance on the interviewees-as-storytellers, the evidence of nonhuman animals' self-awareness and subjectivity in the films inadvertently becomes, at times, contradictory. About five minutes into *Blackfish*, one trainer states that "[w]hen you look into [orcas'] eyes, you know somebody is home. Somebody is looking back. You form a very special relationship with your animal." Half an hour later, another trainer shares her doubts over the nature of her bond with the captive whales:

> At the time I think I could have convinced myself that the relationships we had were built on something stronger than the fact that I am giving the fish. You know, I like to think that. But I don't know that that's the truth.

This echoes Petitto's final claim about her relationship with Nim, when she reinstates the species divide after recalling Nim's violent attack on her: "He wasn't my child. He wasn't my baby. You can't give human nurturing to an animal that could kill you." Petitto's words strike a discordant note in the film's grand attempt to decipher Nim's subjectivity, especially because no in-depth hypothesis as to why Nim attacked her is offered. While the viewer is invited to infer that Nim got violent as a result of once again feeling abandoned by a mother figure, there is no way to ascertain "that that's the truth." It is in these interpretative crevices where questions about the limits of biograph-ability become more noticeable.

As stated earlier, the camera is a hopelessly anthropocentric medium and no true departure from its indexicality of human storytelling can ever be accomplished – only experimented with. This complies with Thomas Nagel's (1974) premise that our phenomenological understanding of what it is like to be another species is limited to the approximation of what it would be like *for a human* to be that other species. As happens in real life, the camera struggles with the (un)knowability of the species other. The attempt to "cover up" that unknowability with the logic of anthropocentrism ought to give us pause. *Blackfish* has received scholarly criticism decrying the coherence of its arguments and the ways it reasons them out. Loretta Rowley and Kevin Johnson regard the film as a perpetrator of "anthropomorphic anthropocentrism" given that it claims the moral value of orca biology, language, intelligence and sociability because they sublimate interspecies communication:

> [O]rcas' aptitude for entertainment work – ironically, the very practice *Blackfish* argues should end – provides the context through which their intelligence is described. Orcas come to be understood as intelligent because they are portrayed as able to understand humans and express a sense of interest and desire to do so.
>
> (2018, 832)

Jason Sperb (2016) also criticizes *Blackfish*'s ambiguity about the reason why Tilikum was triggered to kill Brancheau, given the film's insistence on both nurture (the mental toil of captivity) and nature (the instincts of apex predators) as explanations.

The films' approximation to nonhuman subjectivity begins with the dichotomy between freedom and captivity. Marsh's realistic approach to Nim's circumstances limits the subject's shot at freedom and thriving to same-species company and larger enclosures at a sanctuary. *Blackfish*, on the other hand, is more grounded in the wilderness ideal. Sperb situates the film within the tradition of nature documentary sentimentalization, which affects the representation of wild orcas (2016, 207). Admittedly, the final scene of redemption in which viewers watch the former trainers watch the wild pods is deictically more focused on the interviewees, but there are other images of wild orcas that present themselves differently (though never truly free from anthropocentrism). An early sequence of *Blackfish* features OSHA Expert Witness Dave Duffus describing orcas in these terms:

> The first nation's people and the old fishermen from the coast, they call them blackfish. They're an animal that possesses great spiritual power and they're not to be meddled with. I've spent a lot of time around killer whales, and they're always in charge. I never get out of the boat. I never mess with them. The speed and the power is quite amazing. . . . Even after seeing them thousands of times, you see them and you still [gasp] wake up.

Duffus's words are not so much sentimental as they are mystifying. He likens the spiritual, territorial and visual encounter with orcas to a transcendental experience but appears intent on avoiding anthropomorphism. There is an implicit acknowledgment of their ultimate unknowability, and this is best sublimed with images of wild pods. It is worth noting that the scientific input brought to the film by Howard Garrett (who speaks about orca communication and culture) and Lori Marino (who discusses their highly evolved brain and limbic system) is paired with underwater images of wild orcas, while the ones that aestheticize Duffus's remarks remain above the surface. This visual framing is "truer" to our actual environmental proximity to their species. Without the technology to record them underwater or a glass tank to create a false sense of intimacy, the wild whales paired with Duffus's intervention reaffirm their environmental and visual distance. One catches sight of dorsal fins and blowholes of pods surfacing; these individuals are thriving in their "whaleness," irrespective of whether we can visually contain them or not. By virtue of what can be seen and what the naked human eye, from its terrestrial domain, can behold, the sequence pays its respects to the orcas' wildness. One sees the whales here as nature would have us see them flourish. And those images make absolute sense in the key of Duffus's words because they reveal the type of wonder that, in Nussbaum's view, positively influences an ethical appreciation and encounter with nonhuman animals. According to Nussbaum, wonder is an emotion that is "more active than awe, more connected to curiosity" (2022, 10). She adds that:

> [W]onder is especially closely connected to our awareness of movement and sentience. We see and hear these creatures moving and doing all these things, and we imagine that something is going on inside; it's not sheer random motion, but directed somehow by an inner awareness, by a someone. Wonder is connected to our perception of striving: we see that creatures have a purpose, that the world is meaningful to them in some ways we don't fully understand, and we are curious about that.
>
> (11)

Wonder, in other words, entertains the idea of nonhuman animal subjects being their own biographers, the masters of their own choices to act upon the world. And although, as Nussbaum claims, there is no such thing as "the wild" because there is no place in the world that is not affected by human activity – plus, she adds, "the pretense that 'the wild' exists is a way of avoiding our responsibility" (2022, 224) – freedom from captivity will always provide the more viable scenario in which to move, do and choose.

The images of orcas in tanks in *Blackfish* inspire a different response. There is no denying that the stirring of one's consciousness when beholding images of whales performing tricks or facing, in all their huge magnificence, the glass that separates them from their human onlookers, is guided by the interviewees' accounts of the misery of captive orcas. But there is also the possibility that regardless of their remarks, there is something discordant about watching cetaceans being watched in a spectacle that obliterates their sense of self (a fundamental index of such indignity being their inability to escape and their full body exposure behind the glass, as objects of voyeurism). In a recent study on the effects of music on the emotional response to images of orcas, researchers observed how participants in the study were more likely to perceive the whale as happy if they believed her to be in a wild setting, and sad if they believed her to be held in captivity (Manitzas Hill et al. 2023). There can be no true wonder in captivity – only consumption. And while many would remain skeptical of the notion of happy and sad whales and regard this as anthropomorphism, a greater injustice to nonhuman animals is to deny them complex feelings because these are ascribed to humans only – what Frans de Waal calls "anthropodenial," that is, an "a priori rejection of shared characteristics between humans and animals when in fact they may exist" (1999, 258).

The interviewees in *Project Nim* do not engage in discourses about how chimpanzees thrive in the wild and instead adopt a welfarist approach to devise the most ethical (or realistic) scenario for an adult Nim. However, LaFarge does acknowledge a vital part of Nim's sense of self (his chimpanzee-ness) that was suppressed and thwarted by human interference: "We had done so much damage removing him from what his life should have been. We exploited his human-like nature without regard to his chimpanzee nature. We were co-opting him right from the beginning. It was wrong." Accompanying LaFarge's comments is footage of Nim compulsively pushing a kind of barrel from one end of his cage to the other; a behavior which, in light of LaFarge's statement, is read by the viewer as a stereotypy resulting from captivity and isolation. For all of LaFarge's faults as a maternal figure, her acknowledgment of this nature is refreshing in the context of the shared and implied suggestion from the interviewees that interspecies communication (whether through sign language or other means) is the basis of why Nim should be regarded as a moral subject.

Indeed, the "just like us" approach values certain nonhuman animal species given their similarities with humans and is at the heart of the legal arguments posed by Steven Wise and the Nonhuman Rights Project (NhRP). Wise's *Rattling the Cage* (2000) sought autonomy rights and the recognition of legal personhood for captive chimpanzees and bonobos because they are aware of themselves, they show intelligence, have a theory of mind and can experience complex emotions in ways that are comparable to the experience of being human. Wise eventually expanded this circle to include elephants and cetaceans, grouping great apes and whales in a logic beyond that of being charismatic megafauna. Unsurprisingly, his arguments did not sit well with legal scholars who were against animal rights (Posner 2000). But they were also criticized by scholars of animal ethics for their anthropocentric bias, as only species fortunate enough to be regarded as sufficiently similar to humans would meet the legal personhood requirements, marking a new dividing line between themselves and "inferior" species (Kurki 2021; Gruen and Waldau quoted in Mitra 2015). While, in all fairness, Wise's proposal proves the flexibility of such a line (as the fabrication that it is) and not its rigidity, there is a danger to the overall premise. Nussbaum remains sympathetic yet critical of his views, particularly with regard to how in his book and in *Unlocking the Cage* (Hegedus and Pennebaker 2016), a documentary following the legal procedures initiated by Wise and his team, so much emphasis is placed on the ability of chimpanzees to pick up on sign language.[5] Nussbaum is skeptical about the implications of this skill:

> [U]sing sign language to indicate emotion is something apes do for and to humans, not something they do among themselves. . . . [It] show[s] something about the animal, but not something that lies at the heart of its form of life.
>
> (2022, 33)

In *Project Nim*, LaFarge embraces a similar position when claiming that she did not believe Nim could thrive in the type of methodical sign-language teaching that

Terrace expected. As a woman married to a poet and working for a linguist, she concludes that: "words are a fucking nightmare when it comes to closeness often." *Project Nim*'s interviewees, therefore, are not fully unanimous in their foregrounding of what made Nim a moral subject. As Nussbaum would have it, most of the interviewees' curiosity falls short of actual, ethical wonder; that is, a form of wonder that demands the safeguarding of Nim's "chimpanzeeness" and the acknowledgment of his subjectivity aside from, beyond and despite the human world in which he is forced to live.

Project Nim features consistent images of the chimpanzee gaze "looking back" or beyond and uses them to make the individual stand out, to make him biograph-able. Film can assist with the de-problematization of the multiple ethological discussions about interspecies forms of looking and attempt to answer the question of "what kind of gaze do we want the animal to look back at us with?" (Burt 2002, 42). The cinematic frame can essentialize the meaningfulness of nonhuman animal gazes, even if the science contemplates different hypothetical scenarios. For one thing, studies on looking time that attempt to measure a subject's corresponding eye gaze at visual stimuli can hypothesize about the cognitive processes and perceptions of such a subject, but it is hard to be conclusive about them (Winters et al. 2015). It is also difficult to assess how or whether nonhuman animals perceive camera lenses as visual agents and what the limits of having a theory of mind may be. Furthermore, the fact that while evolutionary and cultural proximity have led some species (namely primates, domestic and companion nonhuman animals and possibly some cetaceans) to develop communicative forms of eye contact with humans, there is ambiguous evidence to suggest that this is the case with other wildlife, although eye-level photographing can help to circumvent these barriers in order to deliver meaningful, focused encounters (Karpan and Karpan 2020).

As Derek Bousé argues in his celebrated study on the close-up shot in wildlife films, the general principle of continuity, which "focuses attention on the action and story, while drawing it away from the editing itself" (2003, 124), is especially important when crafting a narrative about nonhuman animal life. By subsuming close-ups to what the voice-over is explaining or deictically construing them in relation to what is off camera, or in relation to the previous and the following images, nonhuman animal faces and eye contact become meaningful and contribute to the action. Facial close-ups of Nim's gaze, whether in stills or video footage, abound in Marsh's film. For instance, when recalling Nim's attack on LaFarge, Marion Probst, an assistant at Black Beauty Ranch, recalls how:

> I happened to be looking at [Nim] when Stephanie got out of the car. And he saw her and he recognized her right away. And the look on his face was just "Oh *now* you come. *Now* you come. Now [that] I've been through all this. And *now* you come."

These repetitions of "now you come," followed by a single, grave, musical note on strings, are heard over a close-up photograph of Nim looking from inside his cage to something beyond the frame, while the camera slowly zooms in. Nim's palimpsestic demeanor becomes, in this space of continuity, an expression of anger and contempt. Moreover, this strong emotion creates a "false intimacy" (Bousé 2003) between the viewer and Nim that contributes to his individualization as a biograph-able subject. Nim becomes historicized as a sentient being that once felt that emotion in a given time and place.

The close-up shot and the gaze pose additional problems associated with the creation of false intimacy. Bousé notes that "[a] facial close-up can be used to isolate a single animal, to individualize it and to establish a separate sense of identity for it." But, he asks, "*can* a close-up of an animal's face identify it as an individual, given that animals of the same species often seem indistinguishable in appearance?" (2003, 126; emphasis in original). As discussed in Chapter 1, because of own-species bias, humans require considerable time to process nonhuman animal faces (Jakobsen et al. 2021). The closer the phylogenesis, the easier face readability will be. Chimpanzees' physiognomic resemblance with humans certainly allows viewers to familiarize themselves with Nim's countenance and visually make out his features, and so images of his face can effectively be used for continuity. But what about in the case of an orca, whose face is so alien to that of humans (Marino 2014, 22–23) and whose gaze is virtually impossible to make out? I have written elsewhere about the cinematic complications with the orca face due to the species' shape, achromatism, eye position and aquatic environment (Alonso-Recarte 2025). Film has dealt with these phylogenetic differences more or less compellingly by exploiting extreme and detailed close-ups. But even if such shots are used, does the orca gaze still risk being processed not as an index of subjectivity but as a vacant stare given its lack of similarity with the human eye? The fact that *Blackfish* concerns itself much more with full-body images in captivity and the claustrophobic feeling they elicit rather than images of the face or the gaze goes to show that such close-ups are less useful in construing Tilikum (or orcas at large) as traumatized subjects. At the same time, because "face work" is so central to audiovisual biography given its readability (Corner 2002, 99), the faces of the former trainers, by sheer force of on-screen time, verbal communication and own-species bias, pull the narrative in the direction of human trauma.

Marsh's and Cowperthwaite's deliberate cinematographic choices for the representation of their subjects as individuals and as species members are interesting case studies to examine biograph-ability and the anthropomorphic and/or anthropocentric deixis that is at stake in animal advocacy nonfiction. They illustrate the compositional variables that are dependent on species morphology and phylogenetic reliance on the meaningfulness of the gaze beyond Bousé's wildlife film scope. While point-of-view (POV) shots of nonhuman animal characters in wildlife and fictional films have attracted the interest of scholars (Rooks 2016; Cortés Zulueta 2020), the anthropomorphic implications of this shot are often too great for a genre that seeks to proclaim (individual) nonhuman animal subjectivity. More interesting than the POV shot, I argue, is the study of species-specific shot preference in advocacy. Further study on activist filmmakers' approach and exploration of this issue is needed, and it would provide a fine contribution to the wider focus on nonhuman animal representation.

Conclusion

This chapter has sought to analyze a set of strategies employed in biographically oriented animal advocacy documentaries. It has examined the patterns and tensions involved in the construction and control of the narrative that is the nonhuman animal's biography, all of which brings into focus the question of subjectivity and the moral obligations of humans toward it. One of the revealing similarities between *Project Nim* and *Blackfish* is that at the climactic moment when the nonhuman animal "bites back" and attacks the human maternal figure, the interviewees fall into speculation as to what could have gone through the minds of Nim and Tilikum, respectively. Analyzing *Blackfish*, Rowley and Johnson (2018, 836) point to the contrast between the confidence of the interviewees when describing the reasons behind whale-on-whale aggression and their hesitance to provide a satisfactory explanation to Tilikum's attack on Brancheau. In *Project Nim*, the three interviewees recalling Nim's attack on LaFarge (Probst, LaFarge herself and her daughter, Jenny Lee) show different levels of assertiveness. While Probst declares "I think [Nim] was going to kill her," LaFarge and her daughter regard the fact that he did *not* kill her as evidence that there was a conscious intention not to do so. Jenny Lee recalls that in the heat of the moment, there was speculation coming from different witnesses as to what Nim's intentions were. The scenes, which mark the momentum of violent interspecies contact and close the archetypal cycle of the traumatized (captive) subject, may perhaps disappoint those viewers who expected closure regarding motive and intention. Because the interviewees, who have controlled the narrative, cannot produce a final, provable resolution, the films cannot coherently do so either. However, rather than regarding this as a failure, a more ethical response to the issue of nonhuman animals in captivity would be one that focuses on the ultimate unknowability, mystery and alienness that is the mind of these nonhuman subjects. And there emerges the opportunity for wonder to grow and for the viewer to become genuinely curious about what and how life matters to these subjects. Regardless of their level of proximity to humans, every sentient creature has a way of experiencing the world to which they are entitled, and which is determined by their species but also by their individual personality and sense of self. Advocacy insists on the right of these creatures to self-determination and autonomy, that is, to become their own biographers.

Notes

1 *Project Nim*'s film credits refer to a re-enactment unit that includes a primate choreographer, an animatronic chimpanzee (supplied by Animated Extras), a chief puppeteer, a puppeteer dresser and a sculptor.
2 Coincidentally, Marsh's documentary also came out the same year as the reboot of the *Planet of the Apes* franchise with *Rise of the Planet of the Apes* (2011), directed by Rupert Wyatt. This was not lost on film reviewers that compared the two films by addressing the ethics of great ape captivity (Baum 2012; Porter 2013).
3 Cowperthwaite had directed a film about lacrosse in inner-city Denver titled *City Lax: An Urban Lacrosse Story* (2010); and Marsh's previous project had been the Academy-Award winner *Man on Wire* (2008), about high-wire artist Philippe Petit's 1974 stunt at the Twin Towers.

4 Terrace's conclusions have since been refuted by further researchers of primate language learning. For an updated summary of these experiments, see Krause and Beran (2020).

5 A somewhat similar approach is undertaken in the Spanish documentary *Persona [no] humana* (Cuellar and Sánchez 2022), which depicts the legal approach to great ape personhood in the highly publicized cases of the captives Sandra, an orangutan, and Cecilia, a chimpanzee, in Argentina. The film, however, makes it a point to include the research of primatologists who study wild chimpanzees in their natural environment and analyze body language, signs and vocalizations among individuals and groups. This input, however minimal in the film, supplies a counterpoint to the research on human sign language acquisition which carries the weight to support the legal personhood argument.

References

Aaltola, Elisa. 2014. "Animal Suffering: Representations and the Act of Looking." *Anthozoös* 27 (1): 19–31. https://doi.org/10.2752/175303714X13837396326297.

Alonso-Recarte, Claudia. 2025. " 'Black Fish, Black Fish, What Do You See?' Looking at the Faces of Orcas in Animal Advocacy Documentary Film." *Revista Canaria de Estudios Ingleses* 90: 119–36. https://doi.org/10.25145/j.recaesin.2025.90.08.

Alonso-Recarte, Claudia, and Ignacio Ramos-Gay. 2022. "Real Animals and the Problem of Anthropomorphism in Children's Film." In *The Oxford Handbook of Children's Film*, edited by Noel Brown, 408–28. Oxford University Press.

Anderson, Michael, dir. 1977. *Orca*. Famous Films Productions, Paramount Pictures.

Aranceta-Reboredo, Olatz, Leire Morrás-Aranoa, and Núria Almiron. 2025. "Fighting Breeds and Athletes: How Lobby Groups Frame Animal-based Entertainment in Spain." *Perspectivas de la Comunicación* 18: 1–36. https://doi.org/10.56754/0718-4867.2025.3771.

Arnold, Andrea, dir. 2021. *Cow*. IFC Films, BBC Films.

Baum, Bruce. 2012. "Apes, Humans, and Other Animals. Film Review Essay of *Project Nim*, directed by James Marsh, and *Rise of the Planet of the Apes*, directed by Rupert Wyatt." *New Political Science* 34 (1): 111–21. https://doi.org/10.1080/07393148.2012.646033.

Beatson, Peter. 2011. "Mapping Human Animal Relations." In *Theorizing Animals. Rethinking Humanimal Relations*, edited by Nik Taylor and Tania Signal, 21–58. Brill.

Biressi, Anita. 2004. "Inside/Out: Private Trauma and Public Knowledge in True Crime Documentary." *Screen* 45 (4): 401–12. https://doi.org/10.1093/screen/45.4.401.

Boissat, Laure, Laura Thomas-Walters, and Diogo Veríssimo. 2021. "Nature Documentaries as Catalysts for Change: Mapping Out the 'Blackfish Effect.'" *People and Nature* 3: 1179–92. https://doi.org/10.1002/pan3.10221.

Bousé, Derek. 2003. "False Intimacy: Close-ups and Viewer Involvement in Wildlife Films." *Visual Studies* 18 (2): 123–32. https://doi.org/10.1080/14725860310001631994.

Bruck, Jason N. 2024. "The Cetacean Sanctuary: A Sea of Unknowns." *Animals* 14 (2): 335. https://doi.org/10.3390/ani14020335.

Burt, Jonathan. 2002. *Animals in Film*. Reaktion Books.

Callaway, Ewen. 2016. "Scientists Clash over Lifespan of Captive Killer Whales." *Nature* 531 (7595): 426–27. https://doi.org/10.1038/531426a.

Chapman, Jane L. 2009. *Issues in Contemporary Documentary*. Polity Press.

Chattoo, Caty Borum. 2016. "Anatomy of 'The Blackfish Effect.'" *Documentary Magazine*, March 23. https://www.documentary.org/online-feature/anatomy-blackfish-effect.

Colléony, Agathe, Susan Clayton, Denis Couvet, Michel Saint Jaime, and Anne-Caroline Prévot. 2017. "Human Preferences for Species Conservation: Animal Charisma Trumps Endangered Status." *Biological Conservation* 206: 263–69. https://doi.org/10.1016/j.biocon.2016.11.035.

Colling, Sarat, Sean Parson, and Alessandro Arrigoni. 2014. "Until All Are Free. Total Liberation through Revolutionary Decolonization, Groundless Solidarity, and a Relationship Framework." *Counterpoints* 448: 51–73. https://www.jstor.org/stable/42982377.

Corbett, Julia B. 1995. "When Wildlife Make the News: An Analysis of Rural and Urban North-central US Newspapers." *Public Understanding of Science* 4 (4): 397–410. https://doi.org/10.1088/0963-6625/4/4/004.
Corner, John. 1996. *The Art of Record. A Critical Introduction to Documentary*. Manchester University Press.
Corner, John. 2002. "Biography Within the Documentary Frame: A Note." *Framework: The Journal of Cinema and Media* 43 (1): 95–101. https://www.jstor.org/stable/41552316.
Cortés Zulueta, Concepción. 2020. "Cameras that Pose as Animals. Imagining Non-human Animals through POV Shot." In *Outside the Anthropological Machine. Crossing the Human-Animal Divide and Other Exit Strategies*, edited by Chiara Mengozzi, 243–60. Routledge.
Cowperthwaite, Gabriela, dir. 2013. *Blackfish*. Magnolia Pictures.
Cowperthwaite, Gabriela. 2014. "Entrevista con la directora." DVD extra. Included with *Blackfish*. DVD. Dogwoof, Karma Films.
Cowperthwaite, Gabriela, and Tor Myhren, dirs. 2010. *City Lax: An Urban Lacrosse Story*. SnagFilms.
Cuellar, Álex, and Rafa G. Sánchez, dirs. 2022. *Persona [no] humana*. Dacsa Produccions.
de Waal, Frans B. M. 1999. "Anthropomorphism and Anthropodenial: Consistency in Our Thinking about Humans and Other Animals." *Philosophical Topics* 27 (1): 255–80. https://doi.org/10.5840/PHILTOPICS199927122.
Ducarme, Frédéric, Gloria M. Luque, and Franck Courchamp. 2013. "What are 'Charismatic Species' for Conservation Biologists?" *BioSciences Master Reviews*, 10: 1–8.
Dunayer, Joan. 2001. *Animal Equality. Language and Liberation*. Ryce Publishers.
Finn, Stephen Marcus. 2023. *Farmed Animals on Film. A Manifesto for a New Ethic*. Palgrave Macmillan.
Freitas Torres, Denise, Eduardo Silva Oliveira, and Rômulo Romeo Nóbrega Alves. 2018. "Understanding Human-Wildlife Conflicts and Their Implications." In *Ethnozoology. Animals in Our Lives*, edited by Rômulo Romeu Nóbrega Alves and Ulysses Paulino Alburquerque, 421–46. Academic Press.
Fudge, Erica. 2002. "A Left-Handed Blow: Writing the History of Animals." In *Representing Animals*, edited by Nigel Rothfels, 3–18. Indiana University Press.
Hegedus, Chris, and D. A. Pennebaker, dirs. 2016. *Unlocking the Cage*. Pennebaker Hegedus Films, HBO Documentary Films.
Hess, Elizabeth. 2008. *Nim Chimpsky: The Chimp Who Would Be Human*. Bantam Books.
Hosey, Geoff, Vicky Melfi, and Samantha J. Ward. 2020. "Problematic Animals in the Zoo: The Issue of Charismatic Megafauna." In *Problematic Wildlife II*, edited by Francesco Maria Angelici and Lorenzo Rossi, 485–508. Springer.
Howlett, Kate, Ho-Ye Lee, Amelia Jaffé, Matthew Lewis, and Edgar C. Turner. 2023. "Wildlife Documentaries Present a Diverse, But Biased, Portrayal of the Natural World." *People and Nature* 5 (2): 633–44. https://doi.org/10.1002/pan3.10431.
Huggan, Graham. 2017. "Killers: Orcas and Their Followers." *Public Culture* 29 (2.82): 287–309. https://doi.org/10.1215/08992363-3749069.
Jakobsen, Krisztina V., Cassidy White, and Elizabeth A. Simpson. 2021. "General and Own-species Attentional Face Biases." *Attention, Perception, & Psychophysics* 83 (1): 187–98. https://doi.org/10.3758/s13414-020-02132-w.
Javanaud, Katie, Harshmeena Sanghani, and Grace C. Young. 2018. "The Future of SeaWorld." *Journal of Animal Ethics* 8 (2): 133–37. https://doi.org/10.5406/janimalethics.8.2.0133.
Karpan, Robin, and Arlene Karpan. 2020. "Making Eye Contact in Wildlife Photography." *Photo Journeys*, December 29. https://photojourneys.ca/2020/12/making-eye-contact-in-wildlife-photography.
Kossakovsky, Victor, dir. 2020. *Gunda*. Sant & Usant, Louverture Films.
Krause, Mark A., and Michael J. Beran. 2020. "Words Matter: Reflections on Language Projects with Chimpanzees and Their Implications." *American Journal of Primatology* 82 (10): e23187. https://doi.org/10.1002/ajp.23187.

Krebber, André, and Mieke Roscher. 2018. "Introduction: Biographies, Animals and Individuality." In *Animal Biography: Reframing Animal Lives*, edited by André Krebber and Mieke Roscher, 1–15. Palgrave Macmillan.
Kurki, Visa. 2021. "Legal Personhood and Animal Rights." *Journal of Animal Ethics* 11 (1): 47–62. https://doi.org/10.5406/janimalethics.11.1.0047.
Ladino, Jennifer. 2013. "Working with Animals: Regarding Companion Species in Documentary Film." In *Ecocinema Theory and Practice*, edited by Stephen Rust, Salma Monani, and Sean Cubitt, 129–48. Routledge.
Lorimer, Jamie. 2007. "Nonhuman Charisma." *Society and Space* 25 (5): 911–32. https://doi.org/10.1068/d71j.
Louson, Eleanor. 2021. "Performing Authenticity: The Making-of Documentary in Wildlife Film's Blue-Chip Renaissance." *People and Nature* 3 (6): 1147–59. https://doi.org/10.1002/pan3.1028.
Manitzas Hill, Heather M., Elena Svetieva, Sarah Dietrich, et al. 2023. "The Influence of Background Music and Narrative Setting on Anthropomorphic Judgements of Killer Whale (*Orcinus orca*) Emotional States and Subsequent Donation Behavior." *PLoS One* 18 (5): e0282075. https://doi.org/10.1371/journal.pone.0282075.
Marino, Lori. 2014. "Cetacean Captivity." In *The Ethics of Captivity*, edited by Lori Gruen, 22–37. Oxford University Press.
Marsh, James, dir. 2008. *Man on Wire*. Magnolia Films, Discovery Films, BBC.
Marsh, James. 2011a. "James Marsh Talks Project Nim | Empire Magazine." *YouTube* video, 4:13. Posted by *Empire Magazine*, August 10. https://www.youtube.com/watch?v=wAPO3ZN1WQs.
Marsh, James, dir. 2011b. *Project Nim*. BBC Films, UK Film Council, HBO Documentary Films, Red Box Films, Passion Pictures, Roadside Attractions.
Marsh, James. 2011c. "The Q&A: James Marsh, Film Director." *The Economist*, August 22. https://www.economist.com/prospero/2011/08/22/the-q-and-a-james-marsh-film-director.
Massey, Joseph Eric, and Riana Harintsoa Randriamarohaja. 2019. "*Blackfish*: A Crisis of Legitimacy at SeaWorld." *Tourism, Culture & Communication* 19 (3): 195–207. https://doi.org/10.3727/194341419X15554157596100.
Maynard, Lily. 2018. "Media Framing of Zoos and Aquaria: From Conservation to Animal Rights." *Environmental Communication* 12 (2): 177–90. https://doi.org/10.1080/17524032.2017.131074.
Merskin, Debra, and Carrie Freeman. 2024. "Animal Activism Formula for Success. Lessons for Capitalising on the *Blackfish* Effect." In *Animal Activism On and Off Screen*, edited by Claire Parkinson and Lara Herring, 181–208. Sydney University Press.
Mitra, Maureen Nandini. 2015. "Animals Are Persons, Too." *Earth Island Journal* 29 (4). https://www.earthisland.org/journal/index.php/magazine/entry/animals_are_persons_too.
Nagel, Thomas. 1974. "What Is It Like to be a Bat?" *Philosophical Review* 83 (4): 435–50. http://www.jstor.org/stable/2183914.
Naylor, Whitney, and E. C. M. Parsons. 2019. "An International Online Survey on Public Attitudes Toward the Keeping of Whales and Dolphins in Captivity." *Tourism in Marine Environments* 14 (3): 133–42. https://doi.org/10.3727/154427319X15627970573318.
Nichols, Bill. 1983. "The Voice of Documentary." *Film Quarterly* 36 (3): 17–30. https://doi.org/10.2307/3697347.
Nichols, Bill. 1991. *Representing Reality*. Indiana University Press.
Nichols, Bill. 2001. *Introduction to Documentary*. Indiana University Press.
Noske, Barbara. 1997. *Beyond Boundaries. Humans and Animals*. Black Rose.
Nussbaum, Martha C. 2022. *Justice for Animals. Our Collective Responsibility*. Simon & Schuster.
Parsons, E. C. M., and Naomi A. Rose. 2018. "The *Blackfish* Effect: Corporate and Policy Change in the Face of Shifting Public Opinion on Captive Cetaceans." *Tourism in Marine Environments* 13 (2–3): 73–83. https://doi.org/10.3727/154427318X15225564602926.
Porter, Pete. 2013. "Rise of the Consciousness of the Apes. Film Review Essay of *Project Nim*, directed by James Marsh, and *Rise of the Planet of the Apes*, directed by Rupert Wyatt." *Society & Animals* 21 (5): 497–500.

Posner, Richard A. 2000. "Animal Rights. Review of *Rattling the Cage: Toward Legal Rights for Animals* by Steven M. Wise." *Yale Law Journal* 110: 527–41.
Rendell, Luke, and Hal Whitehead. 2001. "Culture in Whales and Dolphins." *Behavioral and Brain Sciences* 24 (2): 309–24. https://doi.org/10.1017/s0140525x0100396x.
Richards, Morgan. 2010. "Realizing Animals at the BBC: New Media Technology and Wildlife Documentary." PhD diss., University of Cambridge. https://doi.org/10.17863/CAM.11681.
Rickard, Diana. 2023. *The New True Crime. How the Rise of Serialized Storytelling is Transforming Innocence*. New York University Press.
Rollin, Bernard E. 1998. "On *Telos* and Genetic Engineering." In *Animal Biotechnology and Ethics*, edited by Alan Holland and Andrew Johnson, 156–71. Chapman & Hall.
Rollin, Bernard E. 2016. *A New Basis for Animal Ethics: Telos and Common Sense*. University of Missouri Press.
Rooks, Isaac. 2016. "What Big Eyes You Have: Animal Point-of-View Shots in Horror and the Limits of Vision." *Spectator* 36 (2): 23–30.
Ross, Stephen R. 2014. "Captive Chimpanzees." In *The Ethics of Captivity*, edited by Lori Gruen, 57–76. Oxford University Press.
Rowley, Loretta, and Kevin A. Johnson. 2018. "Anthropomorphic Anthropocentrism and the Rhetoric of *Blackfish*." *Environmental Communication* 12 (6): 825–39. https://doi.org/10.1080/17524032.2016.1167757.
Ryan, Ephraim, Azwa Hanan, and Ekky Imanjaya. 2023. "Effective Storytelling Techniques in Environmental Documentaries to Promote Lasting Social Change: An Analysis of *Blackfish* and *An Inconvenient Truth*." *E3S Web of Conferences* 426: 1–6. https://doi.org/10.1051/e3sconf/202342601103.
Scott, Karen D. 2010. "Popularizing Science and Nature Programming: The Role of 'Spectacle' in Contemporary Wildlife Documentary." *Journal of Popular Film and Television* 31 (1): 29–35. https://doi.org/10.1080/01956050309602866.
Sontag, Susan. 2004. *Regarding the Pain of Others*. Penguin.
Sperb, Jason. 2016. "From Nihilism to Nostalgia: *Blackfish* (2013) and the Contradictions of the Nature Documentary." *Journal of Popular Film and Television* 44 (4): 206–19. https://doi.org/10.1080/01956051.2016.1161587.
Strager, Hanne. 2023. *The Killer Whale Journals. Our Love and Fear of Orcas*. Johns Hopkins University Press.
Terrace, Herbert. 2011. "Q&A Herbert Terrace: The Interpreter. Interview by Jascha Hoffman, July 14." *Nature* 475 (7355): 173.
Waller, Randall L., and Christina L. Iluzada. 2020. "*Blackfish* and SeaWorld: A Case Study in the Framing of a Crisis." *International Journal of Business Communication* 57 (2): 227–43. https://doi.org/10.1177/2329488419884139.
Wassermann, Sophia N., Edward J. Hind-Ozan, and Julia Seaman. 2018. "Reassessing Public Opinion of Captive Cetacean Attractions with a Photo Elicitation Survey." *PeerJ* 6: e5953. https://doi.org/10.7717/peerj.5953.
White, Thomas I. 2017. "Dolphins, Captivity, and SeaWorld: The Misuse of Science." *Business and Society Review* 122 (1): 119–36. https://doi.org/10.1111/basr.12112.
Wincer, Simon, dir. 1993. *Free Willy*. Warner Brothers, Le Studio Canal +, Regency Enterprises.
Winters, Sandra, Constance Dubuc, and James P. Higham. 2015. "Perspectives: The Looking Time Experimental Paradigm in Studies of Animal Visual Perception and Cognition." *Ethology* 121 (7): 625–40. https://doi.org/10.1111/eth.12378.
Wise, Steven. M. 2000. *Rattling the Cage. Towards Legal Rights for Animals*. Merloyd Lawrence.
Woods, Chelsea L. 2018. "'Are Your Tanks Filled with Orca Tears?': Crisis Frames and Message Convergence in SeaWorld's Tanked Twitter Campaign." *Corporate Reputation Review* 21 (1): 9–21. https://doi.org/10.1057/s41299-017-0039-y.
Wyatt, Rupert, dir. 2011. *Rise of the Planet of the Apes*. Twentieth Century Fox, Chernin Entertainment.
Zimmermann, Tim. 2010. "The Killer in the Pool." *Outside Magazine*, July 30. https://www.outsideonline.com/outdoor-adventure/environment/killer-pool.

3 Looking at Activists

Representations of Human Morality

As films that seek to inspire some form of action on the part of viewers, animal advocacy documentaries often resort to profilmic representations of activists as models of virtue. These rescuers, freedom fighters, abolitionists and reformers are depicted as heroes that follow a moral code worthy of mimesis, thus embodying exemplary behaviors for viewers who feel impotent or at a loss as to how to usefully and meaningfully channel their actions. Núria Almiron et al. (2024) distinguish five major categories in the filmic framing of animal advocates: converts, social justice advocates, grassroots, vegan elite and heroes. These categories are helpful to deconstruct the discourses that are representative of how documentary film creates meaning out of the activist figure. With such considerations in mind, this chapter aims to elaborate on themes and motifs that intersect these categories. To better reflect the concurrence and variety of representations, it undertakes a wider corpus of films (most of them American) than in the other chapters. The first section offers an in-depth analysis of the category of converts, given its status as a motif that is transversal to multiple films. I begin by focusing on the evangelical nuances present in documentaries such as Jenny Stein's *Peaceable Kingdom: The Journey Home* (2009) and Marc Pierschel's *Butenland* (2019), a German film.[1] I then connect this category to what I term "synchronous conversion narratives," which seek a simultaneous transformation of the viewer. In the second section, I examine representations of radical activism and the problematics of anti-terrorist rhetoric in three films: Shannon Keith's *Behind the Mask* (2006), Curt Johnson's *Your Mommy Kills Animals* (2007) and Casey Suchan and Denis Henry Hennelly's more recent *The Animal People* (2019). The three films commit to the principles of animal liberationists and seek to cleanse and redefine their public image, creating a space where the concept of justice is reassessed and measured by moral standards that have proven to be of no relevance for law enforcement or in courtrooms. In the third section, I examine how intersectionality with racial and cultural identity markers affects the representation of activists, and the revisionist criticism deriving from postcolonial sensibilities. I discuss Jasmine C. Perry's *The Invisible Vegan* (2019) as a film that is informed by the politics of intersectionality and as an argumentative exercise that resists rhetorical strategies associated with white privilege. I then consider the criticism surrounding Louie Psihoyos's *The Cove* (2009) to address the perceived shortcomings of the film on the grounds of its white-Western focalization.

DOI: 10.4324/9781032661445-4

The Conversion Narrative

The conversion narrative is a common trope in activist and animal advocacy documentaries. Presented as a journey of transition between opposing positions, it is a useful means to strengthen the legitimacy of the activist cause. As Alexa Trumpy claims,

> credibility does not just stem from the convert's extended previous involvement with the opposition. It comes from the combined knowledge converts gain from their time with the opposition and their current involvement in an organization that campaigns against the opposition.
>
> (2022, 654)

In Chapter 2, I referred to the redeemability of a range of perpetrators who, whether former employees at SeaWorld or scientists who at some point participated in Herbert Terrace's project, confessed their shame and regret for their past actions. In both *Blackfish* and *Project Nim*, these "converts," pushed to the forefront of the narrative as interviewees, sustain the "biograph-ability" of Tilikum and Nim, respectively, and their introspectiveness as authorities on the subject lends weight to the moral message of the films. A host of other films follow suit, offering slight variations on the same theme.

These types of converts were once perpetrators (in)directly engaged in animal exploitation. They might have started off from a place of ignorance, but their journey reveals a growing awareness of their own accountability. Empathy is enabled because they speak from a position of repentance. As such, they inspire forgiveness and supply a face, a visible physiognomy, to the abstraction of reform. Ambivalently, these converts speak from a place of torment (caused by the memory of the suffering inflicted by their hands) and also from a place of peace that arises from their present position, one from which they righteously denounce and reject those previous practices. As overcomers of trials and tribulations, these human characters are presented as phoenixes that befit the (United States) mythology of redeemability, personal renaissance and reinvention of the self (Phillips Casteel 2007; McAdams 2013), all of which are valuable assets in Western meritocracy. These dramatizations draw heavily on evangelical Christian rhetoric, and biographies of activists and advocates might even integrate hagiographic aspects, as can be seen for instance in the portrayal of Dr. Andrew Linzey in *The Animal Thing* (2023), directed by his son, Adam Linzey. Sarah Pike's study of rites of passage associated with radical animal rights and environmental activism discerns the ubiquitousness of Christian language, even at a critical stage of youthful rebellion:

> The notion of a second chance at a new life after a conversion in which one's old ways are left behind is such a pervasive metaphor in American cultural history and American religion that it can lend clout to activist conversion stories.
>
> (2013, 37)

In his study on religious conversion, Lewis Rambo (1993, 41) delineates the social factors that encode the profile of the convert (the more marginal, the more likely to transition), and argues that cultural (in)stability will determine the effectiveness of the conversion: people will be more receptive to conversion in cultures in a state of crisis that makes faults, inadequacies and deficiencies more perceptible. In animal advocacy films that align with the Christian rhetoric of conversion, such as Jenny Stein's *Peaceable Kingdom: The Journey Home* (2009) and Marc Pierschel's *Butenland* (2019), the converts pinpoint the genesis of their transition in the relentless feelings of empathy and compassion that caused them cognitive dissonance. As farmers, this marginalizes them from the ethos shared by the mainstream agrifood community and industry, and they must carve out for themselves a new (interspecies) space of care and coexistence where they can flourish. In communion with their spiritual awakening, they transform their properties into sanctuaries or rescue centers. These topographies are rhetorized as definitive, atemporal re-enactments of the "promised land," and although the films do depict the financial strains and difficulties that the subsistence of sanctuaries might entail (especially in *Butenland*), their conception as spaces that guarantee the happiness, dignity and thriving of farmed nonhuman animals remains unquestioned.

In *Peaceable Kingdom: The Journey Home*, Stein, who in 2000 had already explored the theme of conversion in *The Witness* (discussed in Chapter 1), manages to evoke Christian themes and symbols despite the absence of explicit religious indoctrination. The film collects the accounts and statements of former US farmers and ranchers who transitioned into advocacy. Howard Lyman, known for the scandal he caused in the mid-1990s as a guest on The Oprah Winfrey Show (Oprah openly renounced the consumption of hamburgers after listening to Lyman, and both were sued by the National Cattlemen's Beef Association), Jim Vandersluis and his wife, Cheri Ezell-Vandersluis (owners of Maple Farm Sanctuary) and Harold Brown (also an ex-farmer), among others, are all part of a chorus having experienced the same transformative path. The interviewees are framed as soft-spoken converts (far removed from the stereotype of the radical, in-your-face activist) whose choral narratives trace their path to a new form of existence. In a disquieting juxtaposition with their gentle demeanor, violence pierces its way into the film, either through archival footage or through the imagery evoked by their ekphrastic testimonies.

The camera's patient focus on the countenance of the interviewees while they recount moments of trauma (there will generally be a particular story or experience singled out as a point-of-no-return moment where the farmer can no longer negate or downplay their guilt) becomes an exercise of physiognomic deciphering, whereby the penitent human words and face will deliver the redeemable subject. Everything about *Peaceable Kingdom* echoes the theme of forgiveness: the film's opening image of one-winged angels; Lyman's powerful voice reminiscent of a preacher's congregational speech; the gentle and uplifting musical soundtrack; the occasional slow motion; and the images of cute and caring interspecies interactions when nonhuman animals thrive and flourish in sanctuaries and rescue centers. Cuteness is a definitive strategy for activists and advocates (Lee 2024; Ngai

2023) that can even serve to avert censorship of political commentary (Zuckerman 2015). Cuteness, moreover, inspires interaction and sociality (Maddox 2023, 85) and is connected to the feeling of joy. In line with Jessica Maddox's understanding that "[j]oy is less a response to external stimuli and more a result that comes from actively practicing happiness and existing in specific ways" (2023, 86), Stein depicts the materiality of joy with images of nonhuman animals engaged in caring, physical contact with the converts. The promise of forgiveness is somatized with the sense of touch, with the hand. As Harold Brown says,

> It's like you innately know that this is not right. Otherwise, I wouldn't feel so bad about it. I wouldn't . . . feel so sad. And if it was righteous, if it was a truly good thing, then it would be a joyful thing. But I've never known anybody who slaughters animals and that finds it is a joyful thing.

Through the figure of Brown, Stein draws on another powerful motif in the film: that of the child and what is construed as their innate predisposition toward kindness. Heir to the Victorian and Edwardian domestic ethos of kindness that promoted a form of child-rearing based on evangelical benevolence, fraternity and the caretaking and nurturing practices of the feminine ideal (Grier 2006; Brantz 2007; Traïni 2016, 93–124), the film heavily resorts to notions of spiritual purity and innocence. Lyman's wife, at first shocked by the treatment of nonhuman animals under her husband's feedlot operations, eventually becomes desensitized, only to reconsider her initial response when she observes her children crying at the sight of calves being branded. Brown describes his childhood struggle against an education that encouraged and morally justified human use of nonhuman animals and that encoded normative masculinity in the restraint of empathy. Cognitive dissonance is instilled by carnism – that is, the normalization of a belief system that classifies species into the edible/consumable and the inedible (Joy 2010) – and reveals the widespread prioritization of human taste buds over nonhuman animal suffering (Foer 2009). While, as Hal Herzog (2010) contends, humans can routinely live in moral incongruency and behave in ways that are inconsistent with their feelings of empathy, Stein's conversion narrative finds a fundamental element for transitioning in the cognitive dissonance of children. A key moment of enlightenment is often the realization and subsequent inability to understand why farmed nonhuman animals are treated so differently from companion species. This theme is also prevalent in *Butenland*, when Jan Gerdes, who has inherited his parents' farm, discusses the weight of an educational legacy built on the moral (and economic) justification of exploitation. In the midst of his crisis, he comes across Karin Mück, a vegetarian since childhood, who is transitioning toward less radical and confrontational forms of advocacy. Together, they eventually turn the farm into a sanctuary.

Infancy and innocence are also presented as developmental stages of nonhuman animals that demand moral consideration. In *Peaceable Kingdom*, Cheri Ezell-Vandersluis's tearful remembrance of the young goats that she sold off to sustain the farm comes across as confessions in pursuit of forgiveness (or perhaps even self-punishment). Her verbal description of her sale of a terrified young goat is replete with horrific memories of the brutality with which the buyer placed him in his vehicle and drove him away to a certain death. The ekphrastic horror of Ezell-Vanderluis's words – somewhat reminiscent of Claude Lanzmann's technique in *Shoah* (1985), which abstains from the use of archival footage – serves as a mnemonic token for the viewer to remember the haunting forces of trauma. *Peaceable Kingdom* does show graphic footage of the violence of farm practices, but the personal traumatic stories that the interviewees share are fleshed out only by their words and facial traces, perhaps suggesting that it is the cyclical replaying of their memories that has allowed them to break the wheel and open themselves up to a new form of interspecies community. In this sense, the "journey home" (the title's coda) doubly evokes the convert's Romantic return to the innocent gaze of the child and the return to the promised Garden of Eden as a righteous subject. Herein lies the holy trinity of conversion: from perpetrator to caretaker, from farm to sanctuary/rescue center, and from farmed object to companion nonhuman animal subject.

In *Peaceable Kingdom* and *Butenland*, as in other films such as *Vegan: Everyday Stories* (Lacey 2016), the advocates and activists have already ascended into a superior state of moral awareness. The ritual of confession, of making private guilt public, affirms itself as a mode of expiation, though it is also important to emphasize the peace and joy these subjects experience after transforming their lifestyles and livelihoods. Other documentaries, however, produce "advocates made in real time" (Almiron et al. 2024, 57). I argue that such types of representation ideally aim at the target audience's "synchronous conversion." Rather than focalizing on a convert's verbal reconstruction of the past crisis that triggered the "rebirth," the attempt at synchronous conversion presents converts in the making and invites the viewer to accompany them on their own path to (self-)discovery. A standard formula will involve the same protagonist-narrator taking viewers through his thought process as he brings them along as witnesses to the events and actions marking the logical consecution of the film's objectives. The narrator reconstructs this coherently, using past tenses, while, as a character, he is forever encoded in his present moment, as he pursues his quest for knowledge and truth. Together, protagonist and viewer undertake a journey toward revelation: the illusion of a present moment of "in-the-making," informed by structure and focalization, unveils the discoveries that will bring about transformative change. Unlikely as it may be, these films seek a definitive, synchronous epiphany experienced by both viewer and protagonist that will lead them both to a transition to plant-based diets and humane lifestyles. Surely this is an ideal, as epiphanies leading to absolute, lifelong veganism are rarely prompted by exposure to a single film. In general, "the number of documentaries that have had any demonstrable effect is very small" (Winston 1995, 238). Although the twenty-first century has produced an abundance of animal advocacy films that adopt a variety of persuasive strategies (as this book attempts to show), their definitive impact is hard to measure. Still, in the animal rights movement and in the context of current media overexposure, there is a reliance on documentaries as complementary educational tools, and they are a common instrument in processes of conversion into activism (Pike 2013, 38). As a nod to this potential, synchronous conversion films often operate with an assuredness in their tone and message, as if convinced that information and exposure can *only* lead to proper, moral action: veganism. As a result, some viewers might find these strategies uncomfortably preachy and dogmatic, and shut down to the idea of being lectured with a narrative that is believed to tamper with "the institutional structures of professional ethos of independent documentary" that discourage "media-related 'advocacy' in terms reminiscent of those used for religious proselytizing" (Calvert 2006, 155).

Usually, in these attempts at synchronous conversion, the human protagonists do not have a history of direct violence toward or abuse of nonhuman animals. Instead, they present themselves as average citizens and consumers who, for one reason or another, begin to question the origin of the "absent referents" (Adams 1990) that are routinely consumed by humans in their everyday lives, and take it upon themselves to initiate a quest for knowledge (see Chapter 4 for a deeper theoretical discussion on the absent referent). This might be triggered because they are compelled by some disturbing environmental or animal-ethics-related report or image to do research, or because of a personal experience of interspecies empathy that sparks their curiosity to learn more. The artifice of synchronous conversion is more practical this way: it becomes easier to empathize with the emotional shift of a consumer *in situ* because they provide facial credibility to a space of awe and indignation that is supposedly shared by the viewer.

Moreover, the synchronous conversion narrative draws from the tradition of the filmmaker-as-protagonist (Nichols 1991, 71), which is arguably a masculine creative pattern (see Chapter 4). It also borrows techniques from different modes of documentary filmmaking (Nichols 2001, 33–34): the participatory (where the interaction between filmmaker and subject is pushed to the forefront), the reflexive (attention is called to the act of filmmaking itself, as for instance in scenes where the filmmaker-protagonist describes his plan for filming) and the performative (where the personal affective involvement of the filmmaker-protagonist seeks a response and emotional engagement on the part of the viewer). Charts, graphs and animation are often employed to heighten the entertainment value of factual data; the reliance on objective, scientific data is offered as a counterbalance to affective responses, which are perhaps perceived as being too "irrational" or "feminine." The synchronous conversion narrative featuring a male filmmaker-as-protagonist is employed in films such as Ed Antoja's *Empatía* (2017), Kip Andersen and Keegan Kuhn's *Cowspiracy* (2014), and Mark Devries's *Speciesism: The Movie* (2013). Refreshing exceptions to the male-centered pattern are Jasmine C. Perry's *The Invisible Vegan* (2019), discussed in the third section of this chapter, and Marisa Miller Wolfson's *Vegucated* (2011), in which the filmmaker puts three New Yorkers on a six-week vegan diet while they are educated in the reality behind slaughterhouses, factory farms and the health hazards of meat and dairy. Unlike Antoja, Andersen or Devries, Perry and Wolfson do not seek synchronicity. In front of the camera, they already appear as vegans. Their interest in other humans' responses and in plant-based dieting as a lifestyle (as opposed to the dismantling of a grander capitalist complex) decenters and downplays their profilmic presence in comparison with the screen time of their male counterparts.

In synchronous conversion narratives, viewers follow the hero in his quest not only to bring to public knowledge what, as becomes increasingly evident throughout the films, remains concealed within the walls of exploitation industries, but also to accompany him and partake in his transition from innocence (or naivety) to experience. Sometimes a moment of revelation is presented unequivocally and explicitly, pushing the filmmaker-protagonist-narrator one step closer to anagnorisis. One hour into *Speciesism: The Movie*, and right after establishing the correlations between nonhuman animal exploitation and the Holocaust, the narrator (Devries) exclaims "Oh my God! I saw it!" over images of planet Earth from space, accompanied by ominous dark music. And sometimes the filmmakers provide more wiggle room, such as in the ending of *Empatía*, when a not-fully converted Antoja looks at the camera upon being served a steak as if asking viewers what he should do, and what they would do.

Alexa Weik von Mossner (2021) expresses skepticism toward this format after having tested the effectiveness of *Cowspiracy* on a group of undergraduate students at an Austrian university. *Cowspiracy* is a film that amasses environmental concerns and vegan ethics under US codes that idealize the individual confronting the establishment. Scientific evidence does not hamper the film's Romanticist energy: the filmmaker-protagonist-narrator begins as an environmentally conscious average citizen that complies with green consumer practices believing that his contribution matters and is enough. In other words: he is presented as innocent, until he becomes aware of his gullibility and of the corruptive forces of politics and society. He then sets out to dismantle the meat industry's interests and their clout among organizations that claim an environmentalist agenda. Weik von Mossner (2021, 252) remarks how the foregrounding of a white male perspective can be perceived as problematic and is critical of the filmmakers' reliance on non-peer-reviewed data to substantiate their claim, as it threatens their credibility. And lack of credibility, she concludes, is also an issue when the filmmakers decide to follow the same conversion narrative in their next film. *What the Health* (2017) also stars Andersen "as the gullible everyman who is amazed to find out about the various health benefits of a plant-based diet, demonstrat[ing] that this rhetorical strategy can get problematic when it becomes too formulaic" (Weik von Mossner 2021, 264).

Overall, the conversion narrative offers models worthy of emulation to which the audience can, at least to some extent, relate. Viewers can empathize with feelings of repentance or trauma, or they can identify with a protagonist's sense of planetary responsibility as a common citizen and consumer. Transition is for the most part devised as a personal choice that can lead to righteousness, virtue and a commitment to the value of nonhuman animal life and environmental sustainability. The next section will examine the more extreme end of the spectrum, where the political dimension of personal commitment leads to rhetorical frictions between activism and terrorism.

Radical Activism and Terrorism

In a scene in People for the Ethical Treatment of Animals' (PETA) documentary montage *Unnecessary Fuss* (1984), a smiling young woman stands behind a baboon sitting on her haunches on the experiment table. The baboon, in an obvious state of disorientation and shock, looks on helplessly, occasionally moving her arms uncontrollably. Her shaved head, mapped by sutures. The woman poses behind her, hands on her shoulders, as she jokes around with the people recording her. "You better hope that the anti-vivisection people don't get a hold of this film. . . . They got a nice shot of you," the man recording mocks. A few seconds later, with the baboon unable to hold her head and looking back at the woman, another off-screen laboratory worker continues ridiculing her: "You're gonna rescue me from this, aren't you? Aren't you?" These words are immediately played back over a facial shot of another baboon immobilized by straps, looking at the camera, in an effective editing momentum in which the question becomes directed at the viewer. This brief scene, strategically selected from the sixty hours of videotape that the Animal Liberation Front (ALF) had retrieved from the University of Pennsylvania's Head Injury Clinic and handed to PETA, is integrated toward the end of the 27-minute tape with the purpose of creating a lasting image to haunt the viewer with its polarization of extreme emotional responses: pity and empathy toward the primates versus outrage, anger and disgust at the scientists. Carrie Packwood Freeman and Scott Tulloch describe the scene as "prophetic" (2013, 110), given its inclusion in other activist documentaries as a form of payback and poetic justice.

In a similar vein, a clip included in Shannon Keith's *Behind the Mask* (2006) shows men out on a kangaroo hunt at night. One of them removes a joey from the dead mother's pouch, tosses her on the ground and stamps her to death, as some other man advises, "Better switch your camera off." The fetishistic practice of animal abusers recording themselves while knowingly carrying out some immoral or illegal action makes for good weaponry in the hands of activists who believe in shock tactics, and arguably provokes viewer animosity toward perpetrators to a greater degree than undercover footage. Undercover images can display all forms of raw violence; but perpetrators' triumphal recording of their violent deeds exposes the grotesqueness of their performativity. The refraction in the opposition between perpetrator and viewer becomes more pronounced, precipitating the dissonance between the two moral grounds.

The hostility and rage that viewers can experience when witnessing the violence of undercover footage or of perpetrators filming themselves seems an apt point of departure from which to begin to affectively relate to the abolitionist stance endorsed by groups of activists that are publicly categorized as radical or extreme. These types of activists are characterized by their endorsement of direct action, their demand for urgent and immediate systemic change and their unapologetic tactics to achieve total animal liberation and the collapse of exploitation industries. The three documentaries considered here – Keith's *Behind the Mask* (2006), Curt Johnson's *Your Mommy Kills Animals* (2007) and Casey Suchan and Denis Henry Hennelly's more recent *The Animal People* (2019) – attempt to publicly cleanse the image of radical activists so as to divorce it from the rhetoric that has made their actions prosecutable as terrorist acts. To do so, the films function as platforms in which aboveground radical activists, or those who have become martyrs of the movement by serving time in prison, acquire a "face." They represent themselves as moral subjects with a just cause that can be rationally and affectively defended.

For the most part, the three films rely on a similar type of format and offer a story coherently connected by recurrent themes, historic timelines and strategies common to the activist discourse of animal liberation. *Behind the Mask* and *Your Mommy Kills Animals* portray the counterculture of the movement's different organizations, leaders and grassroots, with particular emphasis on the Animal Liberation Front (ALF) and Stop Huntingdon Animal Cruelty (SHAC), two of the most representative and well-known organizations associated with extremism. Both organizations emerged in Great Britain (in the 1970s and 1999, respectively) and eventually spread to the United States. There is some disagreement as to when the first ALF action took place on American soil (Phelps 2007, 265), although Ingrid Newkirk's *Free the Animals* ([1992] 2022) has for years offered a foundational narrative set in the early 1980s, based on the figure of a heroine under the pseudonym of "Valerie." The ALF promotes direct action "to liberate animals from places of abuse," "inflict economic damage on those who profit from the misery and exploitation of animals," "reveal the horror and atrocities committed against animals behind locked doors" and to perform nonviolent actions by targeting inanimate property and taking "all necessary precautions against harming any animal, human and nonhuman" (Animal Liberation Front 2004, 8). SHAC is formed as a militant response to undercover footage that showed deliberate animal cruelty by employees at Huntingdon Life Sciences (HLS), a large and influential contract animal-testing company that was first based in England and then moved its financial center to the United States. Both the ALF and SHAC follow decentralized organizational structures made up of autonomous cells – a strategy that Donald Liddick (2006, 40) claims to stem from ALF founder Ronnie Lee's mimicking of IRA tactics. Activists engaged in any form of (direct) action that aligns with the organizations' principles can consider themselves members. *The Animal People* reconstructs the story of the SHAC campaigns and the prosecution and conviction in 2006 of six of its members in the United States under the Animal Enterprise Protection Act (AEPA) of 1992, later to be amended by the Animal Enterprise Terrorism Act (AETA) of 2006.

The three films lack a voice-over narrator (although Keith, who is an animal rights attorney, briefly appears as filmmaker and interviewee at the beginning and in the last third of *Behind the Mask*). They resort instead to a structure and rhythm marked by different subthemes, a selection of personal and media stories that fuel emotional responses, and footage covering the abuse of nonhuman animals by different industries (a recurrent clip shows an HLS employee violently shaking and punching a Beagle), as well as rescues and vandalization of laboratories and farms by liberationists. In *Behind the Mask* and *Your Mommy Kills Animals*, these images are often played to the soundtrack of straight edge, ska and punk rock – musical genres that are commonly associated with the ritualized initiation of young people into animal rights extremism (Pike 2013). *Behind the Mask* is probably the most musically informed, as it features Goldfinger frontman and activist John Feldmann as interviewee. It also on occasion resorts to more sentimental accompaniments to live up to the emotional climax of given subplots: the story of the killing of British activist Jill Phipps in 1995 is musically texturized by Stephanie Riggio's "The

Voice Inside (Song of Siddharta)," a song more in the style of singer-songwriter acoustics. Tuning into a counternarrative that subverts the arguments that convicted six of the SHAC 7, *The Animal People* is pulsed by the weight of a heavier musical score reminiscent of political dramas and thrillers. A telling choice on the part of Suchan and Hennelly is to introduce Kevin Kjonaas, one of the convicted activists and former president of SHAC USA, as a Justin Timberlake fan. Beyond the humor and levity brought on by Kjonaas's pop memorabilia, the anecdote foreshadows the film's subsequent deconstruction of the myths and stereotypes that contributed to the imprisonment of the SHAC activists. The point is not to undermine the relevance of the straight edge and punk scenes, but to problematize the public's assumption of the connections between movement leaders and musical genres that allegedly incite violence. Matching this disruption of populist expectations is Kjonaas's declaration that he believes in hierarchy and leadership. This mindset stands far from the anarchist creed of chaos that the US establishment associates with the leftist, political "other" in its discursive manufacturing of enemies of the state.

In the string of interviews used in all three films, Kjonaas (credited as Kevin Jonas in *Behind the Mask*) is one of the faces that is transversal to the documentaries, along with that of fellow SHAC member Josh Harper. *Behind the Mask* and *Your Mommy Kills Animals* also feature interviews and footage of frontline activists such as Rod Coronado, Captain Paul Watson and Jerry Vlasak, among others, who consecutively portray activists as reasonable people with an honorable cause, without abstaining from also showing images of activists publicly yelling and confronting authorities and people at their doorstep. Contrary to the soft-spoken, gentle advocates that appear in the conversion-themed documentaries discussed in the first section, the militancy that moves radical activism is visibly enacted by images of protests, break-ins and vandalism. The films guide the consumption of these images by instructing viewers on the movement's principle that neither verbal harassment nor destruction of inanimate property qualifies as violence (unlike the violence exerted upon the victims of farms and laboratories). There are variables in the tone of the activists that are interviewed: while, for instance, Coronado, Watson and SHAC members use vigorous and assertive wording, others – namely women – appear as storytellers that linger in great emotional despair, such as Nancy Phipps (Jill Phipps's mother) and undercover investigator Michelle Rokke in *Behind the Mask*.

The strings of interviews offer structural support to the films, but more importantly, they supply a lineup of facial close-ups that subvert the image of underground activists homogeneously wearing black bloc and ski masks. Popular imagination associates the underground black-bloc activist with out-of-focus and night-camera optics. They are visually defined by their unsolicited occupation of and trespassing into spaces of exploitation – spaces that are disrupted by their vigorous, corporeal actions. In this way, their bodies are archived into popular imagination by one grand image of the unidentified disturber who is no one and could be anyone at the same time, and whose body is mechanized by movements that discompose the space around it. The subjecthood of the interviewees, on the other hand, is delineated by the nuances of their verbal expression, their countenance and the idiosyncrasies of their physiognomies. As individuals, they are the visibly discernible referents entrusted with the delivery of the moral message that automates the black-bloc bodies.

The image of activists matters greatly, and how the unidentifiability of underground activists affects the public is no moot issue. Dr. Karen Davis, founder of United Poultry Concerns, studied the responses of attendees to a forum on direct action held in 1999. She compared responses to a videotape of ALF activists raiding the University of Minnesota to the reactions to a video showing the direct action of the Action Animal Rescue Team in Australia, where the focus was on the nonhuman animal faces. In the former, "the rescuers look and act like remote, stylized figures rather than flesh-and-blood people, and the animals, including birds and fish, are so far away that it is difficult [to make them out]" (Davis 2004, 207). The ALF video also "shows a posturing, self-centered rescue – despite the anonymity of the rescuers – in which empathy for the victims, however *felt*, is *visibly* lacking" (2004, 207; emphasis in original). Alternatively, the other video not only explicitly focused on the faces of the victims but also made it a point to include the Team's contacting the police and explaining the reason for their rescue. In addition, it showed "the gentleness and firmness of the rescue team (as expressed, for example, by their hands)" (2004, 206). Hands, like faces, offer "immediate readability" and "buttress the notion of universal language" in such a way that they contribute to "global understandability and hence exportability and circulation of film worldwide" (Doane 2024, 563). According to Davis, this particular video inspired the presentation and development of open rescue audiovisual narratives. Open rescues avoid the destruction of property so as not to distract from the main issue (the nonhuman animal victims). In addition, the transparency with which activists admit to their actions and explain their rationale might more readily inspire sympathy on the part of the public (Best and Nocella 2004, 40). Davis applauds certain open rescue videos for:

> overcom[ing] a fundamental difficulty in drawing public attention to the plight of factory-farmed animals: the lack of drama. . . . [W]hen the rescue is visually crafted and deftly narrated . . . you have the drama, the dramatis personae, the tension, a storyline, and a "resolution," in what otherwise appear to be, as in reality it is, a limitless expanse of animal suffering and horror.
> (2004, 209)

The three documentaries analyzed here strive to deliver an image of radical activists that recontextualizes the signifier of the black-bloc body – a body stuck in the dark and grainy resolution of poor-quality footage and informed by anti-terrorist rhetoric. Part of this is accomplished, as argued earlier, by the faces of the aboveground and/or publicly recognizable activists; but there is also the inclusion of footage and stills showing these black-bloc bodies in medium-close-up and close-up shots treating nonhuman animal victims with gentleness and care. These bodily composures are not so much meant to negate the image of the undaunted activist that breaks in and destroys property as to complement and justify it. The system has been designed in such a way that outrage, resistance and rescues are the only viable (and moral) courses of action. The films, therefore, do not build the image of activists from scratch: their portrayals must work with the reassortment and refinement of the intertextual relations that are at play and that have a solid grip on post-9/11 Western imagination.

There is a notable difference between *Behind the Mask* and *Your Mommy Kills Animals*. Whereas the former does not present an articulate view to counter the claims of radical activists, the latter provides more space for contesting positions. David Martosko, from the Center for Consumer Freedom, is featured as interviewee and he agrees with the charges of terrorism that the SHAC 7 faced and attacks their defense based on First Amendment rights protecting freedom of speech. *Your Mommy Kills Animals* is also critical of PETA's pro-euthanasia stance, given allegations that many of the nonhuman animals that are killed are neither sick nor moribund. Both films reveal the people and organizations who refused to be interviewed, with the opening of *Behind the Mask* actually showing Keith's attempt to recruit opposing views, to no avail. The inference behind the lack of cooperation is that these people and organizations have something to hide, without entertaining the idea that perhaps they fear being misrepresented or having their statements spun. *Your Mommy Kills Animals* makes it clear that both PETA and the Humane Society of the United States (HSUS), whose practices are questioned, declined being interviewed. *The Animal People* states that companies such as Stephens Inc. and Marsh Inc., which were targeted by SHAC to cease their connections with HLS, also declined. Except for former HLS Senior Management Darioush Dagdar, whose screentime is comparatively limited, no other spokesperson from HLS (later merged into Envigo) accedes to an interview either. And despite Dagdar's courteous demeanor, "the film paints HLS as a company less interested in preventing animal abuse than making sure the media doesn't find out about what actually goes on in their labs" (Plec 2024, 75).

The few scenes in which an antagonist interviewee (who by definition supports animal exploitation) does appear, nurture the viewer's trust in and desire for objectivity and help to sidestep, at least to some extent, perceptions of the films as activist propaganda. The faces of those who do agree to the interviews bring a visibility that squarely lands the conflict on a dialectic debate between *people* and not just between activists and the ubiquitous, intangible power of institutions. As the people behind the institutions, albeit few, come forward, the activists reciprocate by "removing" the ski masks (even though interviewees are already aboveground or publicly recognizable animalists). While they never actually confront each other as interviewees, the leveling of both parts on the same cinematic grounds nurtures perceptions of objectivity. Nonetheless, the films are far from impartial. The dialogic development of themes and arguments strengthens the illusion of the filmic space as fair ground, but ultimately such force cannot withstand the focalization from the activists' point of view. *The Animal People* provides the SHAC 7 with the opportunity to present their arguments in a way that they were denied during their trial: the defense was not permitted to show evidence of animal cruelty in HLS, thus vanquishing from the courtroom any explanation as to what motivated the defendants' actions, which grossly decontextualized images of their harassment of HLS workers and associates. The interviewees continuously reassert their First Amendment rights, given that the accusation was based on their launch of a website that incited coercion against human targets (and despite the fact that the defendants themselves did not actually commit any of the physical attacks that some HLS workers suffered). Harper's recollections of the trial suggest that their loss was due to a distorted image concocted by the prosecution, framing them as terrorists – a move that, as powerhouse journalist Will Potter insinuates in the film, was part of the government's corporate agenda.

Behind the Mask, *Your Mommy Kills Animals* and *The Animal People* confront the conception of the radical activist as an index attached to the semiotics of the post-9/11 image of the terrorist. The terrorist-as-cultural-construct, nurtured by civil fear and trauma, acquires its twenty-first-century guise with the rhetoric adopted by the Patriot Act, which discursively shifted from the emphasis on freedom to that of national security (Best and Nocella 2004) and broadened the concept of terrorism (Black and Black 2004; Spadaro 2020, 65). Will Potter (2009) contends that these changes, more specifically effected with the passing of the Animal Enterprise Terrorism Act in 2006, gave sufficient room for government authorities to go after aboveground activists. Their prosecution set an example that could deter underground activity, which was already illegal, as well as appease corporate anxieties, and stood as evidence of the sociopolitical success of the so-called "Green Scare" that derived from "the reckless use of the word 'terrorism'" (Potter 2009, 672). As terrorism scholars point out, there is no established, transnational consensus that unequivocally defines what constitutes terrorism and what delimits terrorist acts (Miller et al. 2008; Sorenson 2009; Loadenthal 2017), and the term "ecoterrorism," with its emphasis on property destruction as distinct to violence upon people, has also proven problematic (Carson et al. 2012, 296). Even before 9/11, definitions of terrorism varied on the grounds of intentionality, targets, agenda and imprecise understandings of the notion of violence (Corlett 1996, 164–68).

Behind the Mask and *Your Mommy Kills Animals* were created and released in the tumultuous atmosphere that condensed post-9/11 unrest and escalated toward the passing of the AETA. *The Animal People,* released more than a decade after the SHAC 7 case, does not mitigate feelings of rage and injustice, and Suchan and Hennelly for the most part resist the vectors of temperance that the passing of historical time can often lend to documentary narratives. The SHAC interviewees recall their conviction and the trial with enough nuance to suggest that theirs is a bitter maturity that has been violently forced upon them as a result of the government stripping their innocence as American believers and practitioners of First Amendment rights. Their contempt and distrust in the system are marked by a disbelief at how the events unfolded – a reaction that is construed as a small token of any remaining scraps of their identifiably American purity and innocence. While Kjonaas's unbroken spiritedness (or perhaps the hubris of cockiness) befits the pattern of the fall of the American from innocence to a place where he reasserts and re-empowers his position, Harper's recollections of his fear of the consequences of underground SHAC activism and his *almost* apologetic intimations about the grievances of human targets render him the more sympathetic character. A compelling note is struck when the film highlights the opening of the prosecution: a story of a terrorized child hiding behind the front door to his house with a knife, fearing "the animal people." The film features some of the SHAC 7 activists reading the prosecution's opening, ensuring that the close-ups catch sight of their wincing, hesitation and discomfort. Harper states:

> Obviously I don't want a child to be scared. I don't want a child to be at a point where he's cowering behind the door with a knife. Neither myself or any of my co-defendants had ever protested at Sally Dillenback's house. This was the first of a long line of examples of us being put on trial for other people's actions.

The filmic point of the reading is to provide an updated demonstration of the activists confronting the distortions concocted by the prosecution in order to show their lack of closure and, as the film suggests, their understandable feelings of resentment. "I can't read this out loud. I can't. It disgusts me," SHAC 7 activist Lauren Gazzola interrupts, "[b]ecause it's all smoke and mirrors." Still, Harper's empathy toward the child may be regarded as a Romanticist nod to the preservation of a benevolent (American) innocence, even in the aftermath of what is represented by the documentary as an unjust trial.

The insistence on small details that salvage the interviewees' sense of their own identity as Americans is meant to question the concept of "terrorism" as a signifier latched onto imprecise signifieds that activate an array of frightening images. By problematizing the term and connecting the motivation and tactics of radical activists to the antislavery, anti-segregation and feminist movements (similes that are not only transversal to the three films but also recurrent in classic and current animal rights literature), the filmmakers hark back to the roots of US patriotism. Prior to *The Animal People*, Hennelly had already explored these metaphors in the fictional film *Bold Native* (2010), in which an ALF activist leads an operation across the country while his father, a symbol of the fatherland, tries to locate him before the FBI does (Alonso Recarte 2016). A remark made by Kjonnas in *The Animal People* reverberates across foundational episodes of US history in order to reclaim radical activism as a patriotic act and pull the nation back to a time before the clouding of post-9/11 public sentiment with corporate anti-terrorist rhetoric: "America was founded on agitation, throwing crates at sea into the Boston Harbor because they were going to protest the tea tax. That was economic sabotage. That was direct action. That's not terrorism. That's American." Radical activism, in other words, is a cyclical return to the acts and the attitudes that made America possible. It is an act of love toward the fatherland moved by conscience and directed against the tyranny of the establishment.

The focalization on the activists' perspectives in *Behind the Mask*, *Your Mommy Kills Animals* and *The Animal People* aims, therefore, not only to cleanse the public image of radical activists but to present them as selfless, patriotic figures that ensure the subsistence of the (American) moral cycles that safeguard equality and justice. The films collect the ashes of what remains after corporate and governmental powers are "done" with dissenters, and seek to purge national culture from the distractions and distortions that have immorally instrumentalized post-9/11 trauma.

Looking at Racial and Cultural Otherness

The predominance of white membership in the animal rights movement in the United States has been noted at least since the 1990s. A survey of participants in the March for Animals in 1990, which was considered a landmark protest in the US movement, suggested that more than 90% of the participants were white. "One explanation for the predominantly Caucasian composition of the animal rights movement," Wesley Jamison and William Lunch wrote, "is the socioeconomic differences in income, mobility, education and time availability between the white activists and minorities" (1992, 444). In the 2010s, scholarly studies on the continuing demographic imbalances in the movement turned to the issue of representation and relatability. A. Breeze Harper (2010) criticized the movement's promotional packaging of veganism with images of thin white female bodies at the expense of the representation of nonwhite ones. Claire Jean Kim problematized PETA's controversial instrumentalization of the Holocaust and slavery in some of their exhibits, which gave the appearance "of using or exploiting Jews and blacks in order to sell animal rights to the broader public" (2011, 333). Also questioning PETA tactics, Luis Rodrigues (2019) examined white normativity and the racist implications of the creative choices behind some of their campaigns, which were symptomatic of the organization's disconnection from the social and economic conditions faced by people of color. Corey Lee Wrenn (2016) analyzed the movement's lack of diversity in terms of gender, race and body type. Although mainstream media is hard to control, the movement could be held accountable for their own media, and reforms rooted in social justice and intersectionality were necessary. In a previous article that drew on Harper's theory, Wrenn also established two fundamental reasons why the movement alienated nonwhite communities: "First, the nonhuman movement is framed in a way that normalizes and privileges white experience" (2012), she wrote. This addressed the issue of representation of nonwhite bodies and the salience of fundamental concerns that other communities faced, such as food politics in connection to the colonial experience. "Second, the language and presentation utilized in our outreach excludes underprivileged populations" (2012), by which Wrenn specifically meant that animal rights (academic) literature was readable at a literacy level not necessarily held, on average, by all communities given unequal educational opportunities. Wrenn is on to something here, as this type of literature, which frequently disseminates the correlation between speciesism, racism and sexism (a triad popularized by Peter Singer in his 1975 classic, *Animal Liberation*), can, from a position of privilege, give the impression that animal rights issues are readily equivalent to other social justice causes. Nonetheless, as Will Kymlicka and Sue Donaldson (2014) argued, speciesism and animal rights remained largely marginal and unimportant issues for the progressive Left in comparison to what were neatly classified as human rights concerns. The Left's general consensus that animal rights issues were actually a hindrance was unique, for it was the only struggle dismissed as having a weakening, detrimental effect on other intersecting social justice causes, leaving no proper strategizing to overcome these views.

Ultimately, what these studies confirmed was the general impression that in the United States (and perhaps in the Global North at large) animal advocacy and, especially veganism, were a "white thing" that reflected class and racial privilege while remaining tone-deaf to the social, political and economic circumstances of minorities. Some fairly recent polls have suggested a rise in plant-based dieting in the African American community on account of health benefits and the impact of Black artists and influencers (BBC News 2020). However, more updated and meticulous data is needed to establish whether recent cultural trends and sensibilities have changed these perceptions and demographics as of the mid-2020s.

As films of the 2010s, the documentaries discussed in this final section – Jasmine C. Perry's *The Invisible Vegan* (2019) and Louie Psihoyos's *The Cove* (2009) – deal with the problem of white-centrism in different ways. The former engages with strategies of intersection as a practice of decolonization, while the latter became the target of criticism precisely because of its alleged focalization from the moral perspective of white Westerners.

The Invisible Vegan revolves around the connections between plant-based diets, veganism and the African American community. As mentioned earlier in this chapter, Perry is an example of a female filmmaker-as-protagonist, as she appears onscreen as interviewee (credited as Jasmine Leyva), as convert and as a young Black woman that is a family and community member.[2] The film's tone is down to earth and friendly, as Perry presents a series of arguments in favor of a lifestyle that will lead not only to health benefits but also to a reconnection with West African plant-based diets. This re-historicization situating Black bodies in the genesis of precolonialism is a first step in the reclamation and reappropriation of African American identity. Perry frames her arguments around ideals of authenticity rooted in truc racial heritage – a heritage that has been obscured by colonialism and by the acculturating effects of slavery on Black culinary practices. A bold move on the part of Perry is the criticism of soul food, a type of cuisine made from a variety of meat ingredients, heavy spices and seasoning that derives from the nourishment of slaves in the Southern United States. Katharina Vester qualifies the reclamation of soul food as an example of an "invented tradition, as it presents a strategically simplified narrative of African American cooking that served the goal of unifying diverse and numerous communities into a political (and cultural) whole, evoking a sense of pride and achievement in its members" (2015, 52–53). Despite the 1960s' cultural repackaging and celebration of soul food as part of the "Black Is Beautiful and Black Power ethos" (Miller 2013, 1), perceptions of it being unhealthy have damaged its reputation and possibly affected its lack of visibility at a global scale (Miller 2013, 2–3). However, soul food is still embraced by a large proportion of African American families as a culinary practice that reaffirms community and racial identity. Social gathering around the sharing of food is a dominant motif in Black American culture, and *The Invisible Vegan* reasserts these moments of community-building with photos and recordings of family reunions.

For all the film's feel-good and empowering undertones (including a hip-hop soundtrack with lyrics about plant-based diets), it must be noted that the task Perry sets herself is an extremely complex one, as she strives to become a "visible vegan" by doubly battling the white-centrism of veganism and the African American legacy of soul food. Armed with the input of academics Dr. A. Breeze Harper, Dr. Psyche Williams-Forson, nutritionist Tracey McQuirter and Dr. Milton Mills, MD, the film's reclamation of the Black body begins with the cultural restoration of ancestry in West Africa, where low-fat, mostly plant-based diets seemed to predominate. The quest to return to one's roots, therefore, honors powerful traditions prior to the transatlantic slave trade. With the consideration of soul food as a custom originating in the consumption of the "scraps" and "garbage of the plantation" (to use Mills's term), the film becomes a palimpsest in which to rewrite African American culinary history. At the same time, early in the film, the dominating perception of veganism being exclusionary of nonwhite ethnicities is presented. Within the first three minutes of *The Invisible Vegan*, Perry jokes that her initial reaction to veganism was that it was "that white people stuff" and in one scene a friend quips that a salad "tastes like white people food."

Significantly, *The Invisible Vegan* refuses to acknowledge or entertain analogies that make the systemic exploitation of nonhuman animals akin to the slavery of Black people – a telling choice that reminds viewers, who might have been awaiting such a juxtaposition, of the racial authorship and authority behind the filmmaking. The closest the film comes to addressing the analogy is when a representative of the Food Empowerment Project states the following:

> I acknowledge that we as a movement need to be better about [understanding intersecting issues affecting race and class]. I don't think it means us going into communities of color and saying, "this is what you should and shouldn't do." It doesn't mean comparing slavery to animals and things like that, which I find offensive. I think it's more [about] looking at these structures that have been in place that exploit and kill living beings.

The analogy is not discussed further, although a solid case for the inclusion of animal rights as part of the intersecting issues is made. Distressing footage of nonhuman animal suffering in farms interrupts the positivity of the film's tone, but it is meant to conjoin animal rights with other social justice matters, not to reproduce the analogy. The absence of the discussion raises issues surrounding the (racial/cultural) ownership of certain narratives and the types of discourses and rhetoric employed in them. Note that it is not Perry as a profilmic authority that shuts down the conversation – the comment is made in passing by another of the many interviewees. Behind the scenes, nonetheless, the deliberate creative choice not to thematically engage with the controversy surrounding the analogy is made.

As suggested earlier, the analogy to slavery (and the Holocaust), although historically common in animal liberation parlance and part of the habitus that strengthens activists' sense of communion (Kim 2011), is particularly problematic for its perceived de-singularization, de-specifying, de-contextualization, appropriation and commodification of a community's traumatic history to further another agenda (Kim 2011; Cantens 2024; Polish 2016). As discussed in the previous section, animal liberationists often liken their cause to those of the abolitionists and freedom fighters who forged a distinctively American identity by combatting slavery and tyranny. While groups belonging to other social justice movements have also, not unproblematically, resorted to slavery and the Holocaust as comparable to their own struggles – Kim (2011, 315) lists gays and lesbians, AIDS activists, pro-life activists and Tea Partiers, among others – the species boundary has proven particularly controversial and much more difficult to neatly resolve. Canonical animal advocacy literature supporting the analogies often reinforces the juxtaposition with validation from a prominent member of Black or Jewish communities; that is, the universalization of victimization becomes viable when the activists themselves are part of the comparison group (Cherry 2010, 462–63). Marjorie Spiegel's classic *The Dreaded Comparison: Human and Animal Slavery* (1988) is legitimized to a great extent by author Alice Walker writing the preface, and Charles Patterson's *Eternal Treblinka: Our Treatment of Animals and the Holocaust* (2002) borrows its title from the claim made by Yiddish writer Isaac Bashevis Singer in "The Letter Writer" that "in relation to [nonhuman animals], all people are Nazis; for the animals it is an eternal Treblinka" (1968, 26).

Yet, despite these concessions from members of the Black and Jewish communities, resistance to the analogies remains, and skepticism toward the creative force behind them leads to conversations about convenient cultural appropriation. Slavery and the Holocaust, according to this view, are not to be de-historicized or instrumentalized: given their acquired signification as unequivocal examples of evil and barbarity largely built on the animalization of entire communities and the premise that humanity equates whiteness (Polish 2016), to then speak a rhetoric which blurs the interspecies boundary is to desecrate the ineffability that protects their specificity and an insult to the memory of the victims. Those opposed to the analogies regard the obliteration of the human–animal moral hierarchy as a blasphemy that exposes white commodification of Black and Jewish history and identity. According to this view, the analogies suggest that white culture remains blind to the continuing oppression and violence that these communities face and that are the legacy of the horrors of slavery and the Holocaust.

Perry's creative choice to avoid the analogy as a subject matter or as a rhetorical device for persuasion can, in this sense, be regarded as an act of decolonialization because of the refusal to engage with the type of discourse that is perceived as the product of (white) privilege. Instead, the film proposes and stands by its defense of intersectionality, where animal rights are an important (but not isolated) issue, and where comparability in terms of interspecies suffering is not sustained or naturalized by an equivalence to Black slavery. In this light, the silence regarding the analogy becomes a political act that eliminates and disempowers white-centric vegan rhetoric. *The Invisible Vegan* would seek with this disavowal a rearticulation of the Black communal experience that is not built on a reaction or response to the discourses that have dominated (white) animal advocacy, but that is instead erected on its own terms.

What happens when animal advocacy documentaries trigger suspicions about misrepresentations of human cultural and racial otherness? Depictions of white Western animal activism can become blatantly problematic when set outside the

Global North. Louie Psihoyos's Academy Award-winning *The Cove* (2009), arguably one of the most impactful animal advocacy documentaries to date, follows a group of Western (namely United States) activists and filmmakers as they uncover the economic interests behind the capture and slaughter of dolphins in Taiji, Japan. The film is an impressive and breathtaking feat, packed with an effective dosage of suspense, drama, action and occasional humor, and with a signature use of undercover footage, night vision, military thermal cameras and hydrophones. It is also armed with its own conversion narrative: activist Ric O'Barry, once the dolphin trainer of the original *Flipper* series in the 1960s, describes his own evolution from greed to advocacy as a form of enlightenment triggered by the suicide of Kathy, a captive dolphin.[3] O'Barry, the leading figure in *The Cove*, goes on to appear in later cetacean conservation documentaries such as William Neal's *Long Gone Wild* (2019), often described as the sequel to *Blackfish.*

Implicit in *The Cove* is the conviction that bearing witness will lead to change and the protection of dolphins. The activists' heroic journey consists of the retrieval of graphic, visual and audible evidence of the suffering and painful death of intelligent, self-aware creatures (the film posits an estimate of 23,000 dolphins being secretly slaughtered in Taiji annually). The success of the activists culminates not with their retrieval of such evidence but with its display for others to see. Psihoyos exposes the lies of the Deputy of Fisheries for Japan at the time, Hideki Moronuki. On camera, he shows him images they have collected of the horrific slaughter – a slaughter that Moronuki had just falsely claimed to be quick and humane. In another scene, O'Barry, with a screen strapped onto his chest showing the images, walks into a convention of the heavily criticized International Whaling Commission (IWC) and stands before its representatives before being escorted out. These triumphal performative acts of shaming certainly amount to satisfying climaxes that channel the viewers' growing antagonization and anger toward Japanese fishing and whaling industries. Viewers are meant to feel that by embarrassing defenders of Japanese policies some preliminary form of justice is being served, and the activists' moral superiority and honor is vindicated.

Yet these feelings of satisfaction on the part of the viewer are arguably deceptive. For one thing, images of the slaughter are consumed by film viewers in an entirely different way from the manner that Moronuki and the IWC witness them. Viewers of *The Cove* interpret the images as part of the wider intertext of documentary modes and framing, while Moronuki and the IWC confront the images as raw and isolated material. The film's illusion of the IWC experiencing, by way of shame, some hint of moral shock (discussed in Chapter 1) is in fact inadvertently weakened by a follow-up scene that stages O'Barry and his screen in the streets of Tokyo as hordes of passersby are fast-motioned around him. Some pedestrians stop to look at the screen, but mostly people walk on, suggesting, as Sarah O'Brien (2023, 22) indicates, that public indifference may in fact be the reason behind the continuing slaughter, rather than the agenda of policy makers to keep the killings away from public view. If public indifference is the true reason behind inaction, then the collective feeling of shame is either hindered by apathy or becomes so widespread that it dilutes into meaninglessness. And in connection to this possibility is the question as to whether the footage in and of itself could ever be enough to move significant amounts of people to action. The fact that, as O'Brien (2023, 22) reveals, a year after the film's release Psihoyos chose to mail a copy of *The Cove* to Taiji residents rather than the raw footage of the slaughter, speaks of the awareness of the need to situate these types of distressing images in a contextual narrative that instructs viewers on how they should be (morally) interpreted.

In addition, the performative acts of shaming become questionable tactics when viewed through a postcolonial lens. The whiteness and "Westernness" of the activists, led by Psihoyos and with O'Barry as leading advocate, mark a stark contrast to the Japanese fishermen and the industry that endorses them. One would have to consider whether a Western viewer's perceptions of shame and the belief that the Japanese antagonists are actually *feeling* ashamed (and that their cultural experience of shame is similar to a Westerner's) are not in fact overshadowing other possible emotions that they might be experiencing and that are more aligned with their beliefs and stakes in the slaughter. If such is the case, then the hopefulness experienced by the viewer and the belief in the film as a first step toward action amount to little beyond a false understanding about how speciesist attitudes and practices can change on a global scale.

The Cove's polarization between Western activists and Japanese fishing and whaling communities is heterogenized through forms of representation that some critics have found defacing. Lara Newman's (2015) empirical study of the effect of the film in focus groups evinces that audience sensitization to perceived colonial and racist attitudes problematizes the persuasiveness of the film's desired narrative. James Hatley (2011) criticizes the film's selective identification of perpetrators in a way that negates the faces of Taiji. This is best emblematized by "Private Space," a Japanese man that harasses the activists and screams at them to "go away," and whose depiction is mediated by the voyeuristic viewing on the part of the audience of the activists filming him as he records them back. "In a politics of outrage," Hatley states, "some other is singled out to carry the burden of the wrong, while the impassioned audience, those who are situated behind the pointing finger rather than in front of it, are given leave to insatiate a justice of retribution" (2011, 13). Indeed, in what is possibly the most compelling scene in the film, the camera captures free-diver Mandy-Rae Cruickshank and another activist attempting to get a vantage from the shore of the killing of a group of dolphins. Cruickshank's voice-over explains how the fishermen tried to stop them and butted chests with them. With a broken voice, she then describes how a wounded dolphin managed to swim over the nets until finally sinking to her death. The horrific images of the dying, terrified dolphin are shown, as the activists helplessly look at her and embrace each other, with Cruickshank in tears. In the background of the shot, four Taiji fishermen look at the activists with indifference, with two of them recording these moments of affective intimacy (or perhaps recording Psihoyos's team filming the shot). As the activists and Psihoyos leave, a couple of the fishermen smile and wave them away, perhaps in amusement at what they perceive as a ridiculous spectacle of Western adoration of dolphins, perhaps as part of their indifference, or perhaps for no specific reason that would properly befit the film's polarization. Yet these smiles are targeted by the outraged viewer, prepped to empathize with the dolphin and Cruickshank, whose focalization dominates the scene.

James Hatley (2011, 8–9) says that *The Cove* proposes an aesthetic communion of the (righteous) human body and dolphins by foregrounding the freedivers' underwater movements and silhouettes in a promotional still of the film. But above corporeal kinship, it is the interspecies affective connection that resonates most in the film. This is emphasized by the activists' empathy toward the fatally wounded dolphin and O'Barry's "telepathic" connection with Kathy, the dolphin that triggered his conversion. Conversely, the film depicts Japanese fishermen and whaling industry agents as villains that threaten the sanctity of this interspecies bond. As Melissa Haynes claims in her study of the politics of disgust in *The Cove*, while the film portrays the dolphins as highly intelligent and sentient beings with a rich social life, "the fishermen are shown violently hunting prey, eating raw flesh, and expressing territorial animal rage at the filmmakers' intrusion" (2013, 37). These forms of "animalization" elicit a colonial gaze upon racial and cultural otherness, and the polarization is widened with the film's critique of the IWC delegates of Antigua and Dominica, who are said to be bought off by Japan in exchange for their votes. Such portrayals seem unfair; the demands of moral rectitude from struggling nations overlook the urgent economic, health and (human) social justice issues in these countries. They also implicitly foster the image of the (white) Western savior as a force of civilization who is on a mission to straighten moral codes, bring sustainability back to the oceans *and* save a species as unique as the dolphin. As Carrie Packwood Freeman (2012) ascertains, despite the important animal rights message it inspires, the film fails to critically survey its own Western favoritism for charismatic fauna (at the expense of other "edible" ocean species), and it also overlooks the connections with Western animal agriculture (in fact, O'Barry circumvents the comparison between dolphin and cow meat by almost immediately deviating into the subject of tradition). Freeman also laments the film's missed opportunity to correlate the activists with other social justice movement leaders, such as Harriet Tubman – a type of analogy which, on the other hand and as argued earlier in this section, would be controversial if perceived as an appropriation and instrumentalization of Black history.

Ultimately, what the body of criticism of *The Cove* teaches us, is that despite the admirable task of grassroots activists and filmmakers, and despite the many accomplishments of the documentary that rightfully consolidate its canonical status, the "mark" of white-Western focalization influences narrative form and themes in a way that negates the racial/cultural other's subjecthood, and that reasserts occidental hierarchies of who is worthy of moral consideration.

Conclusion

Animal advocacy documentaries devise representations of reformers, liberationists and activists to provide viewers with positive human models that defy speciesist culture, lifestyles and the animal-industrial complex. Their relatability as subjects liable to evolve, change and meaningfully transform their place in the world renders them as heroic figures that provide viewers with moral guidance. These figures are meant to be admired and celebrated, especially given the trials and tribulations encountered along the path toward a more compassionate form of existence. As should be clear by now, the archetypes and mythical paradigms that structure their quests are identifiably Eurocentric and American. This is unsurprising considering that the corpus analyzed in this chapter is mostly made up of films from the United States, but it has been purposely selected because of their reliance on representative formulas and conventions of the subgenre when it comes to creating a narrative about the spectrum of human morality in a speciesist world. The conversion parable is informed by Christian doctrine, and animal liberation is allegorized as an updated struggle against the tyranny and greed of the establishment and corporate power. The deconstruction of anti-terrorist rhetoric begins with the correlation between Americanness and the activists' freedom of conscience and speech, and the moral imperative to fight injustice. On the other hand, the emphasis on intersectionality proposed by Jasmine Perry exposes the myopia of the white-centric approaches that have traditionally dominated vegan discourses and the animal rights movement. Finally, the criticism faced by a film as significant as *The Cove* demonstrates the need for activists to continue exploring how to gaze upon antagonists and continue questioning and countering their practices without resorting to mechanisms where colonial images and attitudes resonate.

Notes

1 In this chapter, the title *Peaceable Kingdom* is used as a shortened version of the title *Peaceable Kingdom: The Journey Home* (2009) and must not be confused with Tribe of Heart's initial film project, *Peaceable Kingdom* (2004), which was removed from distribution in 2005.

2 Jasmine C. Perry is credited under the name Jasmine Leyva as interviewee and protagonist in *The Invisible Vegan*. To avoid confusion, this chapter will only refer to her with her directorial name.

3 The original hit television show *Flipper* was broadcast by NBC for three seasons from 1964 to 1967. O'Barry participated in the capture and training of the five bottlenose dolphins used in the series.

References

Adams, Carol J. 1990. *The Sexual Politics of Meat. A Feminist-Vegetarian Critical Theory*. Continuum.

Almiron, Núria, Laura Fernández, and Olatz Aranceta-Reboredo. 2024. "Inspiring Animal Liberation. The Representation of Activists in US Animal Advocacy Documentaries (2000–2019)." In *Animal Activism On and Off Screen*, edited by Claire Parkinson and Lara Herring, 39–63. Sydney University Press.

Alonso Recarte, Claudia. 2016. "Animal Liberation, American Anti-terrorist Culture and Denis Hennelly's *Bold Native*." *Critical Studies on Terrorism* 9 (2): 247–68. https://doi.org/10.1080/17539153.2016.1163864.

Andersen, Kip, and Keegan Kuhn, dirs. 2014. *Cowspiracy: The Sustainability Secret*. A. U. M. Films, First Spark Media.

Andersen, Kip, and Keegan Kuhn, dirs. 2017. *What the Health*. A. U. M. Films, First Spark Media.

Animal Liberation Front. 2004. "Animal Liberation Front Guidelines. Reprinted from the ALF Primer." In *Terrorists or Freedom Fighters? Reflections on the Liberation of Animals*, edited by Steven Best and Anthony J. Nocella II, 8. Lantern Books.

Antoja, Ed, dir. 2017. *Empatía*. La Diferencia, FAADA.

BBC News. 2020. "Why Black Americans are More Likely to be Vegan." *BBC News*, September 11. https://www.bbc.com/news/world-us-canada-53787329.

Best, Steven, and Anthony J. Nocella II. 2004. "Introduction. Behind the Mask: Uncovering the Animal Liberation Front." In *Terrorists or Freedom Fighters? Reflections on the Liberation of Animals*, edited by Steven Best and Anthony J. Nocella II, 9–63. Lantern Books.

Black, Jason, and Jennifer Black. 2004. "The Rhetorical 'Terrorist': Implications of the USA Patriot Act on Animal Liberation." In *Terrorists or Freedom Fighters? Reflections on the Liberation of Animals*, edited by Steven Best and Anthony J. Nocella II, 288–99. Lantern Books.

Brantz, Dorothee. 2007. "The Domestication of Empire: Human-Animal Relations at the Intersection of Civilization, Evolution, and Acclimatization in the Nineteenth Century." In *A Cultural History of Animals in the Age of Empire*, edited by Kathleen Kete, 73–93. Berg.

Calvert, Pamela. 2006. "Media and Metanoia: Documentary 'Impact' Through the Lens of Conversion." *Pastoral Psychology* 55: 145–66. https://doi.org/10.1007/s11089-006-0025-2.

Cantens, Kristian. 2024. "Racism, Speciesism, and the Argument from Analogy: A Critique of the Discourse of Animal Liberation." *Journal of Applied Philosophy* 42 (2): 652–67. https://doi.org/10.1111/japp.12780.

Carson, Jennifer Varriale, Gary LaFree, and Laura Dugan. 2012. "Terrorist and Non-Terrorist Criminal Attacks by Radical Environmental and Animal Rights Groups in the United States, 1970–2007." *Terrorism and Political Violence* 24 (2): 295–319. https://doi.org/10.1080/09546553.2011.639416.

Cherry, Elizabeth. 2010. "Shifting Symbolic Boundaries: Cultural Strategies of the Animal Rights Movement." *Sociological Forum* 25 (3): 450–75. https://doi.org/10.1111/j.1573-7861.2010.01191.x.

Corlett, J. Angelo. 1996. "Can Terrorism Be Morally Justified?" *Public Affairs Quarterly* 10 (3): 163–84. https://www.jstor.org/stable/40435951.

Davis, Karen. 2004. "Open Rescues: Putting a Face on the Rescuers and on the Rescued." In *Terrorists or Freedom Fighters? Reflections on the Liberation of Animals*, edited by Steven Best and Anthony J. Nocella II, 202–12. Lantern Books.

Devries, Mark, dir. 2013. *Speciesism: The Movie*. Mark Devries Productions.

Doane, Mary Ann. 2024. "Cinematic Faces and Hands." *Screen* 65 (4): 563–71. https://doi.org/10.1093/screen/hjae046.

Foer, Jonathan Safran. 2009. *Eating Animals*. Little, Brown and Company.

Freeman, Carrie Packwood. 2012. "Fishing for Animal Rights in *The Cove*: A Holistic Approach to Animal Advocacy Documentaries." *Journal for Critical Animal Studies* 10 (1): 104–18. http://www.criticalanimalstudies.org/volume-10-issue-1-2012.

Freeman, Carrie Packwood, and Scott Tulloch. 2013. "Was Blind But Now I See: Animal Liberation Documentaries' Deconstruction of Barriers to Witnessing Injustice." In *Screening Nature. Cinema Beyond Human*, edited by Anat Pick and Guinevere Narraway, 110–26. Berghahn.

Grier, Katherine C. 2006. *Pets in America. A History*. University of North Carolina Press.

Harper, A. Breeze, ed. 2010. *Sistah Vegan. Black Female Vegans Speak on Food, Identity, Health and Society*. Lantern Books.

Hatley, James. 2011. "Blaspheming Humans: Levinasian Politics and *The Cove*." *Environmental Philosophy* 8 (2): 1–22. https://doi.org/10.5840/envirophil20118212.

Haynes, Melissa. 2013. "Regulating Abjection: Disgust, Tolerance, and the Politics of *The Cove*." *English Studies in Canada – ESC* 39 (1): 27–50. https://doi.org/10.1353/esc.2013.0023.

Hennelly, Denis Henry, dir. 2010. *Bold Native*. Open Road Films.

Herzog, Hal. 2010. *Some We Love, Some We Hate, Some We Eat. Why It's So Hard to Think Straight about Animals*. HarperCollins.

Jamison, Wesley V., and William M. Lunch. 1992. "Rights of Animals, Perceptions of Science, and Political Activism: Profile of American Animal Rights Activists." *Science, Technology, & Human Values* 17 (4): 438–58. https://doi.org/10.1177/016224399201700402.

Johnson, Curt, dir. 2007. *Your Mommy Kills Animals*. Indie Genius Productions.

Joy, Melanie. 2010. *Why We Love Dogs, Eat Pigs and Wear Cows. An Introduction to Carnism*. Conari Press.

Keith, Shannon, dir. 2006. *Behind the Mask*. Uncaged Films, ARME.

Kim, Claire Jean. 2011. "Moral Extensionism or Racist Exploitation? The Use of Holocaust and Slavery Analogies in the Animal Liberation Movement." *New Political Science* 33 (3): 311–33. https://doi.org/10.1080/07393148.2011.592021.

Kymlicka, Will, and Sue Donaldson. 2014. "Animal Rights, Multiculturalism, and the Left." *Journal of Social Philosophy* 45 (1): 116–35. https://doi.org/10.1111/josp.12047.

Lacey, Glenn Scott. 2016. *Vegan: Everyday Stories*. Americonic Films, Northwest VEG.

Lanzmann, Claude. 1985. *Shoah*. Les Films Aleph.

Lee, Jin. 2024. " 'Cute Saves the World!': Animal Internet Celebrities' Cute Activism in the #adoptdon'tshop Animal Rescue Campaign on Social Media." *Social Movement Studies*: 1–23. https://doi.org/10.1080/14742837.2024.2421973.

Liddick, Donald R. 2006. *Eco-Terrorism. Radical Environmental and Animal Liberation Movements*. Praeger.

Linzey, Adam, dir. 2023. *The Animal Thing*. Adam and Jesse.

Loadenthal, Michael. 2017. " 'Eco-Terrorism': An Incident-Driven History of Attack (1973–2010)." *Journal for the Study of Radicalism* 11 (2): 1–34. https://doi.org/10.14321/jstudradi.11.2.0001.

Maddox, Jessica. 2023. *The Internet is for Cats. How Animals Shape Our Digital Lives*. Rutgers University Press.

McAdams, Dan P. 2013. *The Redemptive Self. Stories Americans Live By*. Oxford University Press.

Miller, Adrian. 2013. *Soul Food. The Surprising Story of an American Cuisine One Plate at a Time*. University of North Carolina Press.

Miller, DeMond Shondell, Jason David Rivera, and Joel C. Yelin. 2008. "Civil Liberties: The Line Dividing Environmental Protest and Ecoterrorists." *Journal for the Study of Radicalism* 2 (1): 109–23. https://doi.org/10.1353/jsr.2008.0034.

Neal, William, dir. 2019. *Long Gone Wild*. Vision Films.

Newkirk, Ingrid. (1992) 2022 . *Free the Animals. The Amazing True Story of the Animal Liberation Front*. Lantern Publishing & Media.

Newkirk, Ingrid, and Alex Pacheco, prods. 1984. *Unnecessary Fuss*. PETA.

Newman, Lara. 2015. "The Effects of *The Cove* and *Bold Native* on Audience Attitudes Towards Animals." *Animal Studies Journal* 4 (1): 77–98. https://www.uowoajournals.org/asj/article/id/138.

Ngai, Natalie. 2023. "Homemade Pet Celebrities: The Everyday Experience of Micro-celebrity in Promoting the Self and Others." *Celebrity Studies* 14 (4): 437–54. https://doi.org/10.1080/19392397.2022.2070714.

Nichols, Bill. 1991. *Representing Reality*. Indiana University Press.

Nichols, Bill. 2001. *Introduction to Documentary*. Indiana University Press.

O'Brien, Sarah Jane. 2023. *Bits and Pieces: Screening Animal Life and Death*. University of Michigan Press.

Patterson, Charles. 2002. *Eternal Treblinka. Our Treatment of Animals and the Holocaust*. Lantern Books.

Perry, Jasmine C., dir. 2019. *The Invisible Vegan*. Kenny & Jasmine Leyva Films.

Phelps, Norm. 2007. *The Longest Struggle. Animal Advocacy from Pythagoras to PETA*. Lantern Books.

Phillips Casteel, Sara. 2007. *Second Arrivals. Landscape and Belonging in Contemporary Writing of the Americas*. University of Virginia Press.

Pierschel, Marc, dir. 2019. *Butenland.* Blackrabbit Images.
Pike, Sarah M. 2013. "Radical Animal Rights and Environmental Activism as Rites of Passage." *Journal of Ritual Studies* 27 (1): 35–45. https://www.jstor.org/stable/44368863.
Plec, Emily. 2024. "*The Animal People*. Reconstituting Identity in the Animal Liberation Movement." In *Animal Activism On and Off Screen*, edited by Claire Parkinson and Lara Herring, 65–87. Sydney University Press.
Polish, Jennifer. 2016. "Decolonizing Veganism: On Resisting Vegan Whiteness and Racism." In *Critical Perspectives on Veganism*, edited by Jodey Castricano and Rasmus R. Simonsen, 373–91. Palgrave Macmillan.
Potter, Will. 2009. "The Green Scare." *Vermont Law Review* 33 (4): 671–87.
Psihoyos, Louie, dir. 2009. *The Cove*. Ocean Preservation Society, Diamond Docs, Skyfish Films.
Rambo, Lewis R. 1993. *Understanding Religious Conversion*. Yale University Press.
Rodrigues, Luis C. 2019. "White Normativity, Animal Advocacy and PETA's Campaigns." *Ethnicities* 20 (1): 71–90. https://doi.org/10.1177/1468796819873101.
Singer, Isaac Bashevis. 1968. "The Letter Writer." *The New Yorker*, January 13.
Singer, Peter. (1975) 1990. *Animal Liberation*. Avon Books.
Sorenson, John. 2009. "Constructing Terrorists: Propaganda about Animal Rights." *Critical Studies on Terrorism* 2 (2): 237–56. https://doi.org/10.1080/17539150903010715.
Spadaro, Paola Andrea. 2020. "Climate Change, Environmental Terrorism, Eco-Terrorism and Emerging Threats." *Journal of Strategic Security* 13 (4): 58–80. https://doi.org/10.5038/1944-0472.13.4.1863.
Spiegel, Marjorie. 1988. *The Dreaded Comparison: Human and Animal Slavery*. Mirror Books.
Stein, Jenny, dir. 2009. *Peaceable Kingdom: The Journey Home*. Tribe of Heart.
Suchan, Casey, and Denis Henry Hennelly, dirs. 2019. *The Animal People*. Finngate, Seafox, Birdstreet Productions, Virgil Films.
Traïni, Christophe. 2016. *The Animal Rights Struggle: An Essay in Historical Sociology*. Amsterdam University Press.
Trumpy, Alexa. 2022. "Based on a True Story: The Use of Conversion Stories in Social Movements." *Social Movements Studies* 21 (5): 642–58. https://doi.org/10.1080/14742837.2021.1967120.
Vester, Katharina. 2015. "A Date with a Dish: Revisiting Freda De Knight's African American Cuisine." In *Dethroning the Deceitful Pork Chop. Rethinking African American Foodways from Slavery to Obama*, edited by Jennifer Jensen Wallach, 47–60. University of Arkansas Press.
Weik von Mossner, Alexa. 2021. "Feeling Bad? Veganism, Climate Change, and the Rhetoric of *Cowspiracy*." In *Veg(etari)an Arguments in Culture, History and Practice. The V Word*, edited by Cristina Hanganu-Bresch and Kristin Kondrlik, 245–69. Palgrave Macmillan.
Winston, Brian. 1995. *Claiming the Real: The Griersonian Documentary and Its Legitimations*. British Film Institute.
Wolfson, Marisa Miller, dir. 2011. *Vegucated*. FilmBuff, Kind Green Planet.
Wrenn, Corey Lee. 2012. "Deconstructing White Privilege in the Animal Rights Movement." *Animal Rights Zone*, February 6. https://arzone.ning.com/forum/topics/deconstructing-white-privilege-in-the-animal-rights-movement.
Wrenn, Corey Lee. 2016. "An Analysis of Diversity in Nonhuman Animal Rights Media." *Journal of Agricultural and Environmental Ethics* 29 (2): 143–65. https://doi.org/10.1007/s10806-015-9593-4.
Zuckerman, Ethan. 2015. "Cute Cats to the Rescue? Participatory Media and Political Expression." In *From Voice to Influence. Understanding Citizenship in a Digital Age*, edited by Danielle Allen and Jennifer S. Light, 131–54. University of Chicago Press.

4 Animal Advocacy at the Intersection With Gender

The (Fe)male Gaze and the Feminine Aesthetic

The field of Critical Animal Studies has a history of successful coalition with (eco) feminism and animal rights feminism. Particularly since the 1990s, scholars have offered diverse critiques that connect the power relations that objectify women to those that oppress nonhuman animals and vice versa, and have gradually extended their scope to the consideration of masculine and LGBTQ+ identities, while assimilating other intersecting identity markers. This chapter discusses gender paradigms in animal advocacy documentaries, a subject that has received relatively little scholarly attention despite the amount of published critical theory connecting sexism and speciesism. I begin with a critical overview of the methods, arguments and historical and material reasons behind the scholarly approach to animal ethics from a Gender Studies framework, tracing its growth toward intersectionality. In the second section, I examine Laura Mulvey's (1975) concept of the "male gaze" and what Randy Malamud (2012) calls the "human gaze" to address representations of masculinities in animal advocacy documentaries. I argue that although the precepts of Mulvey's paradigmatic "male gaze" are not normative to this documentary subgenre, instances of its variations can occasionally be discerned. The third section analyzes Liz Marshall's *The Ghosts in Our Machine* (2013), Andrea Arnold's *Cow* (2021a) and Victor Kossakovsky's *Gunda* (2020) in relation to Lisa French's (2022) conceptualization of the "female gaze" and the "feminine aesthetic" in documentary film. By comparing the three films, the chapter evaluates the extent to which the female gaze and the feminine aesthetic can deliver identifiable, gendered forms of storytelling and how they affect ethical forms of representing nonhuman otherness.

Gender in the Context of Critical Animal Studies

As anticipated in Chapter 3, the correlation between speciesism, racism and sexism evinces the conceptual framing by many animal activists and scholars of systemic power relations. The rationale behind social and environmental justice concerns is that abusive power relations are rooted in forms of domination that fabricate and perpetuate dichotomous identity categories. These are hierarchized into an unmarked category, representing the acceptable, normative standard, and a marked category, positioned as an inferior deviation from such a norm. Theories

DOI: 10.4324/9781032661445-5

on gender-based processes of "othering" into this marked category are essentialized by the man/woman, male/female and masculine/feminine binarisms, which are mirrored by the culture/nature, man/animal and human/animal divides. These androcentric and anthropocentric correlations and the ways in which they acquire a linguistic, socioeconomic and political dimension have long been discussed by ecofeminists (Merchant 1990; Gaard 1993; Plumwood 1993; King 1995; Warren 1997, 2000; Kheel 2008) and feminist animal rights scholars (Adams 1990, 2003; Adams and Donovan 1995; Gaarder 2011) who expose the systemic subjugation and exploitation of nature, nonhuman animals and women in the historical and material practice of patriarchy. To illustrate this connection, these scholars equip themselves with inventories of iconographies, terms and indexes that conceptually and visually portray women and nonhuman animal as victims of violence.

The revisionism of these references requires the subversion of semiotic practices that are inherent to language, which is identifiably speciesist and sexist and which attests to a deeply ingrained history built on a firm belief in the superiority of humans and man. Joan Dunayer's (1995) seminal work unearths the use of pejorative connotations of signifiers of nonhuman animal species to denigrate women. Calling a woman, a *bitch*, a *cow*, a *hen*, a *chick* or a *fox*, to cite a handful of many examples, indicates not just the oppressor's attempt to control, by way of naming, women's bodies and subjectivities, but also the profound objectification of nonhuman animals themselves. Linguistic practice, Dunayer says, is "even more deeply speciesist than sexist" (1995, 17), and even if her study was mostly reflective of English, similar patterns may be discerned in other languages. While verbal signifiers of nonhuman animals are also employed to describe and (stereo)type men, maleness, other ethnicities or human behaviors, and even if some connotations are intended as complimentary (1995, 16–17), the fact that the majority of terms are used derogatively shows an overall systematization of "othering" crafted by *man*-kind (1995, 19) that subjugates nonhuman animals, women and femaleness alike.

This oppressive language is also the focal point of Carol Adams's monumental *The Sexual Politics of Meat* (1990), which garnered prolific scholarship on the connections between the hunting, butchering, piecing and packaging of meat and the consumption of women's bodies. Women are acculturated as prey-able and symbolically edible matter in heteronormative patriarchies that measure virility against the "weakness" and "passivity" of the animalized female body – a literal and symbolic form of subdual also explored by Jacques Derrida in the schema of carnophallogocentrism (1991, 2008) and by Brian Luke (2007). Pleasure derived from the objectification and annihilation of nonhuman animals (through the mastication and ingestion of their flesh) and women (through fantasies of sexual domination ranging from voyeurism to rape) is fused in disturbing linguistic choices and visual imagery that manufacture normative gender identities of what constitutes a 'proper' boy, man, girl or woman.

The revision of men's epistemologies of rationalism in animal rights theory has also led some scholars to embrace cultural feminism's proclamation of the affective, the emotional and embodied connections (Donovan [1990] 2007). Borrowing from Carol Gilligan's (1982) call for nonviolence and nurture as the concrete forms of responsibility that subjects can morally exercise while experiencing the world-in-relation-to-others, the tradition of feminist care ethics toward nonhuman animals has established a solid identity of its own. Such care ethics celebrate the virtues of relationality and communication, and regard women as being more culturally and naturally predisposed to empathize with other species. According to Velasco Sesma (2017, 112–17), it is the challenge to incorporate these virtues beyond the domestic sphere and into the public domain of social justice, economic sustainability and moral conceptions of principles and rights that limits the care ethics tradition.

One probable outcome of the cultural impact of the fourth wave of feminism, the empowerment of LGBTIQ+ communities, the influence of Critical Race Theory and the revisionism of cis and heteronormative epistemologies on masculinity is that the connection between sexism and speciesism has become diversified beyond the male/female dichotomy to account for the organic manner in which different sexual and gender identities can be physically performed and subjectively experienced. Scholars such as Stacy Alaimo (2010) have noted how less reductive approaches to nonhuman animal sexuality have ignited new debates on the long-standing question of biological determinism versus cultural assimilation. Alaimo contends that the issue of sexual diversity in nonhuman animals matters, and calls for a revision of "[p]redominant modes of social theory . . . which still assume a radical separation of nature and culture [and which] tend to minimize the significance of queer animals" (2010, 56). Biologist Joan Roughgarden (2004) also defies the rigidity of the heteronormative gaze in the field of ethology, with a particular focus on the gaps in natural selection theory. Other scientists such as primatologist Frans DeWaal (2022) have also ventured into the gender debate, albeit cautiously and skeptical of possible ideological inferences. More proactive stances toward animal advocacy have been undertaken by pattrice jones (2011), who describes adopting vegetarianism as much a "full-bodied decision" as ceasing to pass as straight, and who is co-founder of the VINE Sanctuary in Vermont, an LGBTQ-led farmed animal initiative. Sadly, an analysis of documentaries from an LGBTQ+ perspective is left out of this chapter due to a lack of corpus, for although the theme of queer sexualities is being integrated in the genre of wildlife documentaries – most notably, with the Peacock original *Queer Planet* (Watkins 2024) – queerness is still an incipient discussion in animal advocacy documentaries. At the time of finishing this book, Indian filmmaker Harsha Atmakuri's project on Dr. Faraz Harsini, an Iranian scientist, animal activist, refugee and member of the LGBTQ+ community, is still in its stage of principal shoot, but promises to open the discussion in new ways.

Race has also become central to the study of gender and food politics. A. Breeze Harper, editor of the groundbreaking *Sistah Vegan: Black Female Vegans Speak on Food, Identity, Health and Society* (2010), reads veganism (which may be motivated by health issues, animal advocacy or both) as a political act of decolonization whereupon the black body empowers, reaffirms and heals itself (see Chapter 3). Harper sheds light on how systemic racism impedes ethnic minorities from having equal access to healthy foods. She is also critical of the self-created media of the animal rights movement, where thin, white female bodies appear to predominate, leaving Black and fuller bodies underrepresented (and thus, discriminating against them) (Wrenn 2016). Harper's work was recently followed up by *Brotha Vegan: Black Men Speak on Food, Identity, Health and Society* (Adewale 2021), bringing together studies on veganism, Blackness and masculinity.

The links between Critical Animal Studies and Gender Studies, therefore, have grown and evolved toward inclusivity and intersectional understandings of the plurality of identities. However, the impact of this on the actual abolition of animal exploitation practices (or on their transition toward more welfarist standards) is not always so evident. There are some exceptions where gender as an identity marker has arguably been beneficial for nonhuman animals, particularly in cases where the visible face of advocacy has been that of an outstanding woman. One such case is Rachel Carson, whose book *Silent Spring* ([1994] 1962) had a tremendous effect in the United States and abroad, centering the discussion around the public's inadvertent complicity in environmental destruction. Another example is Jane Goodall's pioneering approach to scientific linguistic registers. Goodall is often credited for her perseverance in writing about the chimpanzees she studied in Gombe National Park with gendered binary pronouns (as opposed to "it") (Rees 2007, 882). She also gave them names and used terms such as "emotion" and "personality" to describe their individualities (Morell 1993, 422).

But Goodall's advocacy (along with the conservationist, primatological work of Dian Fossey and Birutė Galdikas) did not represent a direct challenge to the sensibilities and routine practices of Western citizens. Great ape conservation in remote locations of the world is an honorable cause (and exotic affair) in Western imagination, but animal advocacy discourses that target daily practices of Western culture and that "have the nerve" to explicitly connect them to gender oppression have been met with more resistance, to say the least. For one thing, the exploitation by the agrifood industry of the reproductive systems of female nonhuman animals – a topic later discussed in detail – is a prickly issue that has implications for everyday private and public dietary choices. Meat-eaters may feel repelled by vegetarian and vegan arguments if they understand the implication to be that by eating meat and nonhuman animal products, they are being anti-feminist (rather than reading into the systemic sexist nature of certain choices, which refrains from personalizing). Another parallelism that does not always garner public favor is the connection between the suffering of nonhuman animals, women and children in the context of domestic violence, which may be perceived as a trivialization of the human victims. There has been a growth of studies suggesting that wherever violence is unleashed against a (companion) nonhuman animal, there is the likelihood of intrapersonal gender and/or domestic violence taking place (Flynn 2000; Sánchez et al. 2022; Mota-Rojas et al. 2022). Nonetheless, some people may reject the idea that animal cruelty could have moral implications comparable to gender violence or child abuse, believing that this is a disservice to the human victims of intrapersonal violence. Instead, such views limit themselves to regarding animal cruelty as usefully symptomatic of dangerous antisocial behavior. These concerns raise the question of how efficient it is, from an activist perspective, to link animal advocacy messages to feminist ones in mainstream campaigning.

Moreover, despite what for some is a logical parallelism between speciesism and sexism, not all animal advocates level these forms of domination horizontally, and some organizations may even capitalize on sexist objectifications of women's bodies to generate publicity. The publicity stunts and advertisements of People for the Ethical Treatment of Animals (PETA) have long been criticized for advocating compassion while commodifying and exploiting women's bodies through suggestive, sensual and highly erotic visual displays and poses (Deckha 2008; Gaarder 2011, 117–47). There are also other patterns in the animal rights movement that are considered problematic: despite the fact that an overwhelming number of activists are women (Kruse 1999), men have traditionally outranked them as leaders and public speakers (Gaarder 2011, 12). Studies from the 1990s suggest that 68–80% of the members are women (Gaarder 2011, 11), with most of them being white and middle-class (Wrenn 2016, 144). Although there is a need for more updated demographical studies reflecting the impact of current feminism and the late 2010s #MeToo movement on membership, the logic of the pattern is historically consistent with the rise of the humane movement in the nineteenth century.[1] The historical trends in the formation and evolution of the animal rights movement ultimately reveal that issues pertaining to gender do indeed inform and influence anti-speciesist perceptions. Similarly, ethical views on interspecies relations illuminate new understandings of how gender-based systems and dynamics can come into play.

Gendered and "Speciesed" Gazes

With Gender Studies clearly integral to a significant share of Critical Animal Studies scholarship, then the question arises as to how this intersectionality is expressed in the form and content of animal advocacy documentaries. Examining how nonhuman animals and activists are looked at and represented in these films through the lens of gender (and sexuality) allows for a comprehensive exploration of the process of filmmaking as a gendered form of creativity itself.

Considering women's and nonhuman animals' markedness in historical practice, it is helpful to begin by asking how critiques of their representations in film can converge. How the androcentric consumption of the female body translates onto the screen was famously discussed by Laura Mulvey in her foundational essay "Visual Pleasure and Narrative Cinema" (1975), which resorted to psychoanalytic theory to deconstruct the mechanisms that eroticize and fetishize women in Hollywood films. The male gaze is enabled by a scopophilic pleasure that sexualizes the object of desire on the screen (there stands a separation between the observing subject and the onscreen, stimulating object), at the same time that pleasure is sought in the narcissistic impulse to identify with one's perceived likeness – usually the "main controlling figure" (Mulvey 1975, 12) that is embodied by the male protagonist. This desire to control and to provide (male) audiences with sources of pleasure results in the perpetuation of women's "*to-be-looked-at-ness*": they emerge as objects of display that are constitutive of the world's "sexual imbalance," where "pleasure in looking has been split between active/male and passive/female" (1975, 11).

Although neither nonfiction film nor nonhuman animals fell within Mulvey's scope, her thesis echoes speciesist forms of visual consumption. Here, Carol Adams's concept of the "absent referent" is a helpful tool for interpretation. Absent referents, Adams explains, are literal and symbolic means where an "original meaning [is] undercut as it is absorbed into a human-centered hierarchy" (1990, 42). Nonhuman animals are transformed into absent referents when they become meat: their living bodies and selves are replaced by dead bodies and body parts, and the gastronomic language that humans employ erases the linguistic trace back to a living being. Furthermore, they "become metaphors for describing people's experiences" (Adams 1990, 42), pulling them away from their subjecthood and subjectivity. The notion of the absent referent is comparable to stereotyped and formulaic representations of women's bodies as stylized through the male gaze. Female characters that are the object of sexual desire are symbolically exhibited as comestibles and consumables; they become "pieces of meat" to be devoured and sexually ravaged. These male-centered fantasies divide the female body in a number of ways (often using shots that direct the focus onto fetishized or sensuous body parts) and render female subjectivity absent, leaving behind a body that conforms to male desires.

Randy Malamud recodes Mulvey's male gaze into the "human gaze" that is representative of how nonhuman animals are indexed in visual culture:

> The phenomenon of looking at animals in visual culture is predicated upon the assumption that the viewer is human and the object is animal. . . . The practice of consuming visual culture embodies an unbridled omniscient lust ensuring the visual object's absolute subalternity. The animal is rendered vulnerable, free for the taking, in whatever way the human viewer chooses.
>
> (2012, 74)

Indeed, nonhuman animals are spectacularized in ways that highlight the agency of the observer: in the same way that women are cast in stereotypical roles arousing and stimulating the fantasies of the male gaze, on-screen nonhuman animals likewise become cuddly sidekicks, charismatic tragic figures, melodramatic heroes, instruments of comic relief or unsung villains. Nonhuman animals are systematically instrumentalized in ways that satisfy the expectations of the human gaze and, at the same time, reproduce the male gaze that is syntagmatic of sexist structures. Nonhuman animals bear their "*to-be-looked-at-ness*" too: their formidable bodies magnetize the eye in the spaces in which humans try to frame them. By becoming proxies for human attitudes and meaning-making, they also intersect with heteronormative expectations. For instance, in mainstream children's film, nonhuman animals often serve as coming-of-age tokens for boys and girls to graduate into adult manhood and womanhood (Alonso-Recarte and Ramos-Gay 2022). In horror and slasher films, animality is often associated with deviant male killers using nonhuman animal body parts and remains as part of their domestic décor (Clover 1992, 24, 26, 32, 47). In hit television sitcoms and soap operas that have, albeit only slightly, dealt with the controversy of animal experimentation, it is the female character (versus the male professor, surgeon or scientist) that is ethically troubled by vivisection. In this vein, shows such as *Good Morning, Miss Bliss* (Brinckerhoff 1988), *Beverly Hills, 90210* (1993–1994) and *Grey's Anatomy* (Frawley 2008) have created episode plotlines or small arcs for their empathetic female characters to express their concerns for the nonhuman animals used in the procedures. Conversely, in American and British crime series, the role of the prominent activist is generally reserved for males, while, in German television, women seem to more adequately fit these roles (Roscher 2024, 31). In nonfiction productions, particularly in wildlife documentaries, the shielded, heteronormative representations of nonhuman animal sexuality play a safe hand for mass media consumers (Mills 2012). Overall, mainstream filmic and television entertainment have tended toward representations of nonhuman animals that strengthen gender values and sexual expectations, particularly regarding the man/woman and masculine/feminine divides.

When it comes to the profilmic representation of human masculinities, animal advocacy documentaries can reproduce (eco)feminist claims of the oppression of nature by hegemonic masculinity. Because men tend to outnumber women as workers in slaughterhouses, factory farms and fishing boats as well as in blood sports, they more readily visually embody the agents of violence, triggering moral shock (see Chapter 1). Since a high percentage of these men belong to lower-class, non-white ethnicities, the racial and colonial implications of Western industrialism cannot be ignored (see Chapter 3). As Gwen Hunnicutt contends, "[t]he killing of animals becomes racialized when a value distinction is made between 'sadistic' individuals who kill animals in ritualized performances, versus 'civilized' folk, who still kill animals, but do so through an elaborate industrial apparatus" (2020, 71). This covertness of the Global North's whiteness is often a subject in documentaries that revolve around the themes of industrial profitability and its conspiracies – themes which, incidentally, are often tackled by male directors or co-directors. Expository documentaries that break down capitalist and political stakes, such as Robert Kenner's *Food, Inc.* (2008), or Giulia Innocenzi and Pablo D'Ambrosi's *Food for Profit* (2024), also effectively associate (white) industrial decision-making with aggressive masculine corporativism. These representations adhere to the sociohistorical construction of maleness and manhood dependent on the exploitation of gender, species and racial others. Here, the erotic of power, meat-eating, rituals of violence and financial and professional success become intertwined with predatorial practices that are encouraged and perpetuated by patriarchal systems (Luke 2007; Hunnicutt 2020).

The criticism of oppressive masculinity in animal advocacy documentaries often relies on showing the abundance of male agents involved in the animal-industrial complex. Although industrial stakeholders are frequently identified, corporativism becomes diffused in labyrinthic scales of competencies and responsibilities, and the mechanical acts of factory farming and slaughterhouse laboring impede a perception of workers on an individualized level. When visible, the faces of these factory and slaughterhouse workers are the faces of many; their bodies, indistinguishable from other male automata. Although they visibly appear as perpetrators, they are often also perceived as victims of a system that treats them as replaceable commodities subjected to inhumane labor. In these cases, they lose their status as victims when the violence they inflict is visibly done with malice, rather than detachment and impassivity.

Nonetheless, animal advocacy films also provide examples of positive masculinities that promote ethical veganism, caretaking and nurturing. Such films can move beyond ecomasculinity's overpowering focus on ecology and the environmental crisis to more directly address the relations between men and nonhuman animals – a much needed shift in studies on ecomasculinity (Twine 2024). These films tend to individualize their male human protagonists, some of whom may be the filmmakers themselves. The level of intimacy (created with close-ups, observational shots of their routines, interviews, voice-overs, etc.) often plays well with their personal stories of transformation and growth into advocacy, reflecting their own crisis of masculinity as one that has allowed them to transcend normative gender expectations. Louie Psihoyos's *The Cove* (2009), Jenny Stein's *The Witness* (2000) and *Peaceable Kingdom: The Journey Home* (2009), Ed Antoja's *Empatía* (2017), Yeray López Portillo's *Yo Galgo* (2018) and Pippa Ehrlich and James Reed's *My Octopus Teacher* (2020), among others, explore the cathartic shedding of masculine dominance and provide a glimpse into what a world of constructive masculinities could look like for nonhuman animals. It should be noted, nonetheless, that all such models of masculinities are white and with limited insight into the pluralities of masculinities or genders, which precipitates the need to introduce other intersecting markers of identity to broaden the notion of ecomasculinity (Gaard 2022).

Returning to the notion of the gendered and "speciesed" gaze, in the context of animal advocacy documentaries, given their *raison d'être*, the attempt to represent nonhuman animals as sentient subjects would, in theory, preclude the absolutism of Malamud's "human gaze" or Mulvey's "male gaze." This prospect, however, is not always so neatly defined. A fine example of this is Louie Psihoyos's *The Game Changers* (2018), a film that persuasively tracks the connections between veganism and plant-based dieting, healthy nutrition, fitness and top-level athletic performance by relying on a glorifying visual display of vigorous muscles and sweating bodies excelling in strength, agility and vitality. *The Game Changers* is a celebration of "hegan" culture that patches together the stories of athletes, gym-goers, and physical trainers whose transition into veganism explains an exponential improvement in their performance and overall well-being. In line with monomythic patterns, the transition (into veganism) and the wisdom that comes with it represent the final stage of a personal quest, in which the hero acquires some object or knowledge that allows him to transcend his previous self and return "home" (symbolized by training, intense workout or competition). Although the film also purposely portrays the journey of female athletes, Joseph Campbell's ([1949] 1968) unequivocal path of the archetypal monomyth as a traditionally masculine pursuit automates an androcentric encoding in the form of storytelling – an unsurprising move considering the triumvirate leading the film's panel of executive producers: James Cameron, Arnold Schwarzenegger and Jackie Chan. This complies with the subject matter itself: heganism, a term coined by Kathleen Pierce in a 2010 article in the *Boston Globe* that combines the masculine pronoun "he" and "veganism." Heganism describes the cultural trend of a hypermasculinized lifestyle based on healthy exercise and, most importantly, the exclusion of animal protein from one's diet. Heganism has not gone unnoticed by scholars who have pointed out that for all of its implicit critique of the association between meat-eating and hegemonic masculinity, the phenomenon still markets a type of normative masculinity based on a muscular physique fueled by nourishment, exercise, male-centered rationalism and self-control, and the mythical hero that defies the establishment. "Sticking it to the man" is reconfigured in hegan discourse as a rebellion against fast-food culture and the mainstream's image of the carnivorous "real man" (Wright 2015; Greenebaum and Dexter 2018; Randall 2018; Jones 2021). In addition, heganism does little to relieve male vegans from their stigmatic image as weak, effeminate and emasculated. As Thomas Randall notes, "it appears that rather than making veganism more acceptable to men, heganism instead creates a new dichotomy between two masculine identities: the hegans and 'unmanly vegans'; the latter identity here continues to be ostracized" (2018, 226).

Admittedly, *The Game Changers* does not adhere to Mulvey's central concern about the objectification of women per se, but its structural and thematic approaches, along with what it makes (or does not make) of the nonhuman animals that are implicated in these dietary choices, manifest a variation of the male gaze that reaffirms hegemonic precepts. One of the most surprising aspects of the film is its lack of explicit commitment to animal advocacy, as there is very little commentary on how veganism sabotages the interests of industries that farm nonhuman animals for food and constitutes a form of care (Carretero González 2025). Rebecca Jones states that while plant-based diets are "frequently more focused on broader environmental concerns and personal health and improvement," veganism "requires engagement with the ethical question of animal exploitation" (2021, 106). Following this definition, although it may seem reasonable to categorize *The Game Changers* as a documentary that revolves around plant-based diets, Psihoyos's directorial résumé, with films such as *The Cove* (2009) and *Racing Extinction* (2015) to his name, makes the absence of explicit animal advocacy a striking choice. As Jones (2021, 111) suggests, the film may intentionally circumvent animal ethics discourses to avert associations between the type of masculinity it promotes and the "effeminizing" acts of care and nurture that such discourses frequently uphold. This results in what Margarita Carretero González (2023) calls the "paradox of the absent referent," whereby Vegan Studies and culture sidestep the issue of nonhuman animal suffering to focus on health and environmental criteria, thus reinstating a rationale that circles back to the discursive obliteration of nonhuman animals in speciesist language. Ultimately, *The Game Changers* exemplifies what a variation of the male gaze can look like in vegan documentaries. This gaze does not rest on the scopophilic objectification of women, but ironically replaces nonhuman animals' "*to-be-looked-at-ness*" with their absence, thus favoring a hegemonic discourse of hypermasculinity.

Deciphering the Female Gaze and the Feminine Aesthetic

In her work *The Female Gaze in Documentary Film* (2022), Lisa French seeks to define and describe what she identifies as a form of documentary filmmaking that reflects the experiential singularity of female directors, affecting formalistic and cinematographic choices as much as the selection of themes, approaches and focalization. Far from proclaiming that there is a monolithic or single aesthetic of the female gaze, French insists on the need to address the plurality and diversity of such gazes, where individual experiences and creativities across the globe fuel a cross-cultural network that is nurtured by a heterogeneity of perspectives and meaning-making. Wary of reductionisms, she presents a vivid assemblage of this plurality while simultaneously critically surveying how women have navigated the politics of the industry, how they verbally assess their productions, and how their films arguably stand as examples of the female gaze. French points to the possible shortcomings of the proclamation of such a thing as the female gaze, particularly regarding the resistance and concern that some documentarians may have about the labeling of their work and being pigeonholed as female filmmakers, as well as the danger of overlooking intersectional forms of identity. She also refrains from making any claim that these women regard their work as representative of the female experience; the filmmakers' preference and understanding of where their work stands as a historically bound intertextual signifier is respected and acknowledged. Nonetheless, by tracing and historicizing creative and formalistic tendencies that seem to recur in women's nonfiction films and that different filmmakers rely on for storytelling, she convincingly sets forth the theory that female subjectivity strongly influences their filmmaking. For this reason, the female gaze should not be viewed as an inverse of Mulvey's male gaze (French 2022, 53), but as a concept that exposes critics and audiences to the multiplicity and diversification of female perceptions, experiences and subjectivities.

Bearing this emphasis on plurality in mind but also acknowledging French's suggestions as to what could qualify as illustrative of a female gaze, this section examines Liz Marshall's *The Ghosts in Our Machine* (2013), Andrea Arnold's *Cow* (2021a) and Victor Kossakovsky's *Gunda* (2020). The fact that a film by a male director is included runs contrary to French's criteria for defining a corpus, as her understanding of the female gaze requires "a focus on how female directors have expressed themselves as people who identify as female" (2022, 6). However, French also refers to "a feminine aesthetic," by which she means a "style" that is detachable from the experiential subjectivity that is constitutive of the female gaze. The feminine aesthetic "is not innate or essential to women filmmakers" and is useful for the purpose of "describ[ing] characteristics, or distinctive and identifiable forms" (2022, 77). French draws on Julia Knight's claims that the feminine aesthetic delimits a way of looking that explores "*how* their protagonists experience or perceive things" and how such experience is represented (Knight 1992, 122–23; emphasis in original). I contend that *Gunda*, for the most part, merits being discussed as a film that exalts many elements that are proper to the feminine aesthetic; but it also contains features that, as we will see, can be perceived as characteristic of male subjectivity. I also see Kossakovsky's film as a rare exception in the range of animal advocacy documentaries made by men, which, as a corpus, are generally characterized by a faster rhythm, a more frequent use of narrators guiding the interpretation, a penchant for dramatic suspense and the targeting of entire businesses or industries on a grand scale. The three films, I argue, comply with a form of representation of the plight of nonhuman animals where elements of the female gaze and/or the feminine aesthetic are discernible.

The Ghosts in Our Machine follows Canadian photojournalist and author Jo-Anne McArthur in her quest to document the deplorable conditions systematized by the animal-industrial complex and her struggles with a sympathetic yet reluctant publishing world to get her images front and center. While the film begins *in media res* (McArthur has, by that time, ample experience in the field of animal photojournalism, a term which she herself coins), the structure still resonates within the monomythic pattern of the heroic quest, during which intimate portrayals of nonhuman animal rescues are offered – among them, a cow named Fanny, her calf Sonny, and two beagles, Maggie and Abbey. Danial Azhar describes McArthur's missions as "espionage-like," for her activism "is one that is covert rather than retaliatory, interventionist or aggressive" (2021, 23). The layering of visual storytelling is significant in the film, as McArthur's photographs are integrated into the moving images of Marshall's shooting. McArthur herself is never visually positioned as an interviewee, but voice-overs of her own provide insight into her subjectivity, and the genesis of her quest is revealed when she recalls her childhood friendship with hens.

McArthur never looks at or addresses the camera, thus strengthening the fly-on-the-wall effect that magnifies the viewer's sense of intimacy with the purity of a heroine who is repeatedly authenticated by virtue of the ways in which she honestly engages in conversation with other humans, gently soothes nonhuman animals and carries herself in the spaces she occupies. McArthur's moments of silence, which are many, are paradoxically loud: Marshall's patience and attention to detail in moments of quiet intimacy are reflective of a restless mind that, paralleling her geographical itineration, is always gearing, both creatively and as an advocate. Marshall captures the subtlety of McArthur's wincing, closing her eyes or taking a deep breath to recompose herself in the presence of nonhuman animal suffering. Marshall, who acknowledges the influence that third-wave feminism had in her initiation into filmmaking (2024, 211), describes her craft as "open-ended, more philosophical and less polemical, not fast-paced" (2024, 213). She notes how she is "drawn to the interiority of the protagonist, as this layer of storytelling guides audiences to personally identify. Interiority is expressed through close-ups, pacing, sound or music, and prioritizing personal detail" (2024, 213). These "psychological interiors or interiority" (French 2022, 66) are common in women's documentaries. Lorena Elke Dobbie, who participated as a consultant and researcher in the film, recalls her interest in "formulat[ing] ideas around the animal question and feminism" (2024, 139) and in bringing different theorists into the shaping of an adequate perspective for *The Ghosts in Our Machine.* She suggests that Marshall's emphasis on the affective and on the emotional explains her cinematic lingering on spaces that spotlight nonhuman animals as subjects.

Precisely because of the centrality of nonhuman animal individuals, *The Ghosts in Our Machine* at no time risks becoming a documentary that is exclusively about McArthur. Contrary to, for instance, Matthew Galkin's *I Am an Animal: The Story of Ingrid Newkirk and PETA* (2007), in which the PETA co-founder and president is the clear nucleus in the conflict (the sole A story), Marshall's female gaze places emphasis on relationality (as opposed to claiming a single protagonist as the thematic crux). By prioritizing the protagonist's relationships with other nonhuman animals and humans, focalization is dispersed, and the personal stories of the nonhuman animals are organically pulled toward the center of the storyline. While, admittedly, McArthur's quest functions as the structural link, the stories of these nonhuman animals are not portrayed as distant or peripheral; on the contrary, these subjects emerge as irreplaceable individuals that push the themes of empathy and interconnectedness to the forefront. This reflects Marshall's own self-defined style as "character-driven" (2024, 212). To that effect, the film's enduring close-ups of nonhuman animals (whether it be the happy excitement of dogs, or the terrified eyes of foxes at fur farms or of pigs being transported to the slaughterhouse) rest heavily on the imperative to recognize the emotions the subjects are experiencing in that very moment. One particularly illustrative example of the manifest focus on nonhuman individuals is a shot of Sonny, the calf, lying and nestling his small, bony body on the hay as he is being transported to the Cornell University Hospital for Animals. With the camera at an eye-level to his, he looks back at the lens as he rests his head. His vulnerability is evinced by his lack of movement, perhaps due to sickness or exhaustion, and his angular frame attests to his fragility. Diagonally behind him, in the left corner, is McArthur, caringly glancing back at him. The film, in other words, should not be viewed as having a central A story and more minor B and C ones, but as stories in constant intersection and relation to those of others. Here, formalistically and thematically, Sonny's story becomes the focus as both viewer and McArthur rest their gaze on his body.

The fact that Marshall opts to center her narrative on a woman activist and nonhuman animal victims can be regarded as a gesture proper of the female gaze. Drawing on Stella Hockenhull (2017, 60), French (2022, 22) claims that women directors tend to provision their subjects (who are frequently "othered," stereotyped or dismissed as inferior in hegemonic culture) with the space in which to freely exercise their voice. This is frequently associated with the choice women filmmakers make not to appear in the profilmic scene or to act as narrators. In animal advocacy documentaries, there are numerous instances of male directors following the trend of "filmmaker as protagonist" (Nichols 1991, 71) set by the likes of Ross McElwee and Michael Moore in the 1980s (French 2022, 22). As discussed in Chapter 3, these filmmaker-protagonists often appear in the guise of heroic discoverers unveiling the concealed secrets of industries. Some examples include Psihoyos in *The Cove* (2009), Ali Tabrizi in *Seaspiracy* (2021), Rob Stewart in *Sharkwater* (2006) and *Sharkwater Extinction* (2018), Kip Andersen in *Cowspiracy: The Sustainability Secret* (2014) and *What the Health* (2017), Mark Devries in *Speciesism: The Movie* (2013) and Morgan Spurlock in *Super Size Me* (2004).[2] Sometimes this presence of the male director is more visually elusive, but still powerful enough to command focalization. This is exemplified by Werner Herzog in *Grizzly Man* (2005), which does not strictly qualify as an animal advocacy documentary, but instead uses the fascinating figure of advocate Timothy Treadwell as a pretext for the director-narrator-character to explore and negotiate his own understanding of nature and the impenetrability of nonhuman animal subjectivity (Sheehan 2008, 117–18; Pick 2011, 151–83; Hediger 2012).

Conversely, Victor Kossakovsky and Andrea Arnold, like Marshall, abstain from appearing onscreen in their works. If anything, in the case of *The Ghosts in Our Machine* and *Cow*, the use of handheld cameras may inadvertently signal their presence, since the bumpiness and wobbliness of movement, along with the improvisational handling when subjects block the view, serve as reminders of the act of filmmaking. *Gunda*, on the other hand, builds its delicate and intimate storyline with the Steadicam of cinematographer Egil Håskjold, who succeeds in keeping the bulky equipment at an eye level with the nonhuman animal subjects. This does not mean that the trope of "unveiling" is missing; quite the reverse, it remains overwhelmingly pervasive, as the films become revelations of the subjective experience of nonhuman animals. While *The Ghosts in Our Machine* insists on the act of seeing by using the motif of visual mediums such as photography and filmmaking, *Gunda* and *Cow* resort to a tactical use of cinematography that implodes ordinary conceptions of farm spaces, offering instead alternative perceptions of time, movement and physicality that arouse a somatic response to what it *feels like* to be the living, manufactured nonhuman animal.

As in Marshall's film, the protagonists in *Gunda* and *Cow* are "othered" subjects: Gunda (which in Scandinavian means "female warrior"), is a sow kept at a humane, free-range farm in Norway,[3] and Luma is a cow living on an English dairy farm. The interest in a limited number of subjects allows for a more dedicated and intimate portrayal of subjectivity and individual sensorial experience, as opposed to the principle of quantity (see Chapter 1) applied in men's films in the line of *Earthlings* (Monson 2005), *Dominion* (Delforce 2018) or *Lucent* (Delforce 2014). However, and unlike *The Ghosts in Our Machine*, the films' focalizations completely refrain from offering a human perspective. In *Gunda*, humans are permanently off-screen, with only their constructions and mechanical instruments as remnants of their uncanny presence – a characteristic that inspires Benjamín Schultz-Figueroa (2022) to read the documentary in the key of the science fiction genre. In *Cow*, farmworkers and their instruments are integral to the space that demarcates Luma's mechanized existence, but the film's lack of interest in their focalization manages to, at the same time, keep them on the sidelines. One of the most perturbing effects of their presence is that, when gathering what little we can absorb about these workers, many of whom are women, it becomes hard for the viewer to categorize them within the archetype of the villain despite an overwhelming impulse to do so after being pulled into the miserable experience of the main character. Contrary to the numerous films that show disturbing footage of violent, aggressive (and usually male) farmhands and slaughterhouse workers, the farmers in *Cow* continue with their everyday routines degrading nonhuman animals with calmness and serenity. Their handling of Luma involves remarks that hint at their awareness of cows as sentient beings (such as a soothing "Good girl!" or a celebratory "You're free!") and this causes cognitive dissonance in the sensitized viewer. These welfare-based forms of handling, more optimal for both the mental well-being of nonhuman animals and for production efficiency (as less confrontational behavior saves time and energy), become terribly unsettling as the viewer struggles to comprehend how, if one is aware of the nonhuman animal's subjectivity, one is still capable of enslavement. What Arnold may be suggesting with this revision of polarization (see Chapter 3) is that humans in fact live with that incongruency every day, and most are completely untroubled by it. By representing farmers in this light, *Cow* provides a mirror to reflect our own complicity as consumers despite the overwhelming evidence of cows' sentience.

Gunda and *Cow* revolve around the life of a mother, a parallelism that has been noted by some scholars who have grouped the films together in their analyses (Schultz-Figueroa 2022; Reijnen 2023; Porter 2023; Duarte and Mailat Gurghian 2025). *Cow* was shot in the span of four years (plus another four for production), delivering the highlights behind the manufacturing of bovine motherhood. Gunda's piglets are two months old by the film's ending. The motif of motherhood is significant here. Not only are women more likely than men to create stories about female subjects and their subjectivities, which are marked by their sex and gender (French 2022, 29–30), but mothers and motherhood often emerge as subjects of interest in their work as well. As French writes, "[f]ilms by men about mothers tend to claim a place in history and the achievements of the legacy of their mothers, or speak to mothers more broadly . . . rather than to focus on relationships with mothers and family stories" (2022, 62). Motherhood and the sexual violence that enables farming is also a recurring topic in animal rights feminism: female nonhuman animals are cyclically impregnated as they are forced into being mounted or artificially inseminated, a process that anthropocentric societies and lawmaking refrain from associating with rape culture (Cusack 2013). These practices are integral to "the cycle of objectification, fragmentation, and consumption" that make the absent referent possible, and that linguistically and conceptually regard nonhuman animal bodies as "food-producing unit[s]," "protein harvester[s]," "converting machine[s]," "crops" and "biomachines" (Adams 1990, 47). *Gunda* and *Cow* are structured around farming cycles, beginning *ab ovo* with the motif of birth (Gunda's piglets seem just days old, and Luma's calf is being pulled out from inside her) and progressively circulating toward a tragic ending that pushes the viewer into anagnorisis insofar as one recognizes oneself as complicit to the perpetuation of the cycles. The endings represent a convergence between the exceptionalism that the viewer has come to bestow on Gunda and Luma (as the films' protagonists) and the knowledge that theirs is the grim experience of millions of farmed nonhuman animals every day. Gunda's offspring are loaded onto a truck (presumably off to slaughter), and the long, final sequence follows the bewildered and anxious sow as she looks for them in the farmyard. In *Cow*, Luma meets her final fate once her body has been spent after repeated impregnations, and she is unceremoniously killed by a farmer with a bolt gun. Prior to her premature death, we have witnessed Luma being mounted and giving birth to two calves, both of whom are almost immediately removed from her side and taken to outdoor crates where they are fed via artificial teats. Luma's teats, in the meantime, are plugged into tentacular milking machines while standing in the rotary milking parlor, the perfect mechanicalness of which stuns viewers into a surreal sense of unease. In this way, Arnold's representation of lactation reflects the scholarly analysis of the dairy industry as a system of colonization of the nonhuman animal and of *all* the mothers who produce breast milk for their children (Gaard 2013; Cohen 2017).

In all three films, spaces are essential to the motif of unveiling, and they are inextricably linked to the type of focalization that distinguishes the female gaze and the feminine aesthetic from normative forms of documentaries employed by male directors. In *The Ghosts in Our Machine*, Marshall's formidable attention to rhythm is instrumental for the representation of spaces: McArthur seems to be forever on the move, traveling to Europe, New York City and Toronto, sharing with viewers the textural and sonic landscapes of planes, kitchens, fur farms, zoos, the hustle of urban geographies, the pastoral peace of Farm Sanctuary in New York and the warmth and quiet of the rooms where she writes and works on the photographs (the "goods" that she retrieves in her quest). Marshall's reconstructions of spaces strategically play with detail and focus. For example, in a scene toward the beginning of the film that depicts New York City streets, a contrast is made between the rushed busyness of its traffic and inhabitants and the many indexes of nonhuman animal products that, to a desensitized eye, go unnoticed. The result is a collage that for a few moments draws on the principle of quantity (see Chapter 1) with the indexing of body parts and absent referents. Attempting to represent what McArthur perceives when walking the streets, Marshall's montage shows images that range from the weary legs of a carriage horse and a woman in the crowd sporting a fur coat, to the windows of retail businesses, butchers' and cosmetic stores – all of which depend on nonhuman animal bodies in some way. Here McArthur's and Marshall's subjectivities visually resist the masculine propulsion of public spaces, especially of the type that celebrate the hecticness of capitalist machinery, and decelerate toward a paused sequence of focal points.

Another important space in *The Ghosts in Our Machine* is the kitchen. As McArthur joins friends and fellow advocates for home meals, detailed shots of appetizing vegan cuisine not only stir a hunger for taste but also become integral to the idea of community-building safe in a domesticity walled from the mechanical world of animal exploitation. Ingrid Newkirk regards kitchens as strategic places of female control that are "fonts of power, the rooms in which plans can be laid for dietary domination. Women can control not only what they eat but also what their families eat" (2011, 70). The female gaze is frequently drawn to such "domestic environments or private spheres, particularly the family" (French 2022, 66). Through the depiction of McArthur's personal intra-human and interspecies relationships, Marshall suggests that vegan community-building lays the foundation for empathetic and compassionate familial structures.

French (2022, 58) also suggests that a characteristic of women's documentaries may be the connection between space, duration and length of shot. This is mostly evident in *Gunda*, where there are variances between longer shots where the subjects appear "being themselves," in the full presence of their existence, and more up-close, detailed shots that strengthen the sense of communion with the somatic experience of the body. Comparatively speaking, *Gunda*'s setting is idyllic: not only do Gunda and her piglets roam at ease in their barn and in the fields but there are also other nonhuman animals whose state of happiness is portrayed. Most notably, the film captures with cinematographic finesse how a group of visibly abused chickens respond when, for the first time in their lives, they set foot on a field outdoors. Their exploratory eyes and their slow steps as they feel the texture of the ground and hear the sound of the wind, insects and birds, create an incredibly sensorial and subjective perception of being in the world. The diegetic sounds that the film foregrounds contribute to such patience, oblivious to the rushed pacing of the world beyond. In the climactic ending, however, as Gunda searches for her babies, the pastoral setting reveals itself for what it always was: a farm that produces meat. The spell is broken, and the intimacy that the viewer has experienced by being made privy to Gunda's motherhood enables the decoding of her snorts as laments of desperation. Space shifts away from a focus on the present and replaces it with a nostalgia for the past and an anxiety about the future. The theme becomes one of *Ubi sunt* – not of those long ago but of those that were just here, in the present. In this way, *Gunda* is transformed into an elegy for all absent referents, and Gunda becomes an unexpected symbol of a pietà whose mourning will never receive closure.

Spaces in *Cow*, on the other hand, are treated with a different kind of patience. As the cycle of milk production moves incessantly, Arnold seeks to dedicate time to Luma's response to her circumstances. Unlike Gunda, and except for her brief release into a grazing field, Luma appears deprived of all agency. Her subjectivity is decentered by the multiple ways in which she suffers indignity after indignity. The world outside, referenced by celestial planes, a hot air balloon, fireworks and flocks of geese, rolls on in indifference, and their recurrence as motifs keeps the story in an uneasy and perplexing apprehension of contrasts. Particularly captivating in the construction of spaces is Arnold's use of music, some of which is the diegetic repertoire of pop radio that plays in the cow shed, and some of which was edited in as part of the montage. Most of the tunes feature female or non-binary vocalists: Billie Eilish, Soak, Charlotte Day Wilson, Kali Uchis, Olivia Dean, Kelsey Lu and Mabel, among others, texturize a soundscape of youthful angst with lyrics revolving around romantic struggles, personal insecurities, depression or sexual longing – themes and emotions often associated with coming of age into adulthood (and womanhood). In Luma's case, growth and exploitation as a female eventually lead to her execution. "I thought it was so interesting," Arnold says, "because pop radio is often songs about love, about longing, about desire – you feel like there's a longing in the cow shed, and desire, for relationships not had" (2022). Kathleen Vernon refers to music as the "most highly gendered and specifically 'feminine' of the different components of cinema and sound" (2017, 47) given its historically grounded aesthetic association with the emotional and the passions of the irrational, which have rendered it a useful tool to ascertain the male gaze. Filmmakers and critics alike have tried to disengage and liberate female characters from musical constrictions. One way to go about this is by compiling a score and/

or soundtrack that invites audiences to bring in their own knowledge and personal experience with the songs along with their perceptions of the musical genre at hand (Vernon 2017, 50). I argue that this is what takes place in *Cow*: Arnold uses pop music hits of the time that the viewer is likely familiar with, recontextualizing them in an alternative setting of capitalist forces of production and consumption. The irony of the contrast between the gendered perspective of commercial music (manufactured at studios) and the objectification of Luma's reproductive system outlines a space that is doubly satirical and disheartening. During the final credits, the bleakness brought on by the song "Milk," by the band Garbage, incorporates Luma's death "into a post-grunge aesthetic of numbness and ironic remove" (Schultz-Figueroa 2022, 52). Lead singer Shirley Manson sets the final tone of the film with the darkness of lyrics devised with simple sentences such as "I am milk / I am red hot kitchen," "I am lost / So I am cruel / But I'd be love and sweetness / If I had you," and the repeated line in the chorus, "I'm waiting, I'm waiting for you."

Another feature that the three films share (and which again returns to the issue of focalization) is the emphasis on the eye of the nonhuman animal – the soulful bridge to subjecthood. The eye becomes an instrument with which to aesthetically demand an ethical regard of nonhuman animals not as a props, stand-ins or proxies of humans, but as subjects with their own interests and personalities who deserve forms of representation that specifically attend to their consciousness – what Josephine Donovan (2011) aptly calls "the animal standpoint." Through the representation of the eye(s), one can distinguish directorial preferences that align with the female gaze and/or feminine aesthetic by virtue of their emphasis on the relational and the empathetic. These preferences can be indicative of the level of directorial compromise with abolitionism. Of the three films, *The Ghosts in Our Machine* more clearly follows an activist agenda on account of its subject matter and messaging. As an activist film, close-ups into the nonhuman animals' eyes looking back at the photographic or film camera prevail. "By capturing the animals' gaze," John Drew writes, "Marshall confronts [the] culturally-ascribed absence and forces the viewer to accept their presence, fleeting though it is. . . . [T]he central emphasis on animals' gazes creates a visual platform from which they communicate" (2016, 210, 211).

Arnold and Kossakovsky, on the other hand, have indicated that their intention was not necessarily to create activist documentaries in the strict sense of the term, although these claims are nuanced by their statements about their reasons for making the films. In an interview for *The Film Stage*, Arnold suggests that the film will likely arouse thinking about meat consumption, sustainability and nonhuman animal exploitation, but she is verbally cautious about delivering a more straightforward and instructive answer: "I want [*Cow*] to be received however people want to receive it . . . I'm going to let everyone have their own relationship with it" she states. When asked if she is a vegetarian or vegan, she says, "I'm not going to talk about that because I don't want it to become about that" (2021b). These hesitations shed light on the film's own stylistic consistency with the motif of Luma's eyes. "I realized early on that her eyes were everything," Arnold says, "so I wanted to keep her head in the frame as much as possible" (2021b). Shots of Luma looking back are recurrent and durable, but her plight is somewhat guided by the film's eloquent penchant for representing the disconcerting and the bewildering with more readily identifiable realist techniques. Over-the-shoulder shots are extremely recurrent as well, and they are intelligently used to direct the viewer's gaze to where Luma is looking. One such frame has Luma, from the indoor pen, looking outside toward the crates where calves are taken (hers having just been removed from her side). The shot plays with the shift from out-of-focus to focus on the crates, a vision made even more disturbing by the contrast between maternal suffering and the arcadian sunniness of the outdoors.

Meanwhile, Kossakovsky's reservations about filmic activism can partly explain why his message of compassion is built on cinematographic subtleness and delicacy. Although Kossakovsky claims that "it's a time for Gunda," and that "we need [*sic*] revolution of empathy," he refrains from qualifying his film as what he calls "vegan propaganda." "I am not politician [*sic*] . . . . I cannot make propaganda, because it's against art" (2021a), he states. I would argue that his decision not to treat nonhuman animals' gaze back to the camera with the same level of patience that he applies elsewhere is also a choice made to avoid what he views as potentially propagandistic. Eyes are a window to subjectivity, but rather than seeking empathy through the leitmotif of nonhuman animals looking back, there is a focus on representing them as more and beyond what they may convey when their eyes meet ours. Kossakovsky's choice of filming in black and white not only allows for a heightened appreciation of textures, contours and shapes; it is also a powerful mechanism to reflect (and reflect on) the subject's gaze, which is mostly directed at the surrounding environment and other nonhuman animals. Because of the sharpened visibility of their sclera, iris and pupil, Gunda's eyes strongly resemble that of humans. Empathy is key, but when it comes to the gaze, it is an empathy that does not seek to affirm itself by drawing on the immediacy of the interspecies affective response (the nonhuman animal looking back, inquiring what is to become of her), but instead opts for a more contemplative understanding of the subject's right to be, irrespective of her awareness or consciousness about human presence – a presence which, as stated earlier, is aesthetically construed using visual metonymies of instruments, constructions and equipment as stand-ins for otherwise absent humans.

Kossakovsky's belief in the subject's right to be, away from "propagandistic" meddling, involves other formal aspects such as the omission of a narrator or voice-over guiding viewers' interpretations, just like in Arnold's *Cow.* But a more problematic choice made by Kossakovsky is the decision to show a scene in which Gunda steps on the runt of her litter, crushing and killing the piglet. As Kossakovsky explains,

> Gunda kills her kid, and the first reaction was, "Let's not use it." But then I said, "Wait a second, how [do] we not use it? Who are you to judge? She's lived here many years before you. She knew he will not make it – he didn't run to her nipple, he was not doing well."
>
> (quoted in Grierson 2021)

Kossakovsky argues that keeping the scene is proof of his creative choice to swerve away from vegan propaganda and implies that the lack of sentimentality in his portrayal of Gunda is more truthful to her subjectivity: "I'm not the one who can judge her. . . . She knows what she does" (2021b). The ethical debate as to human intervention to relieve nonhuman animals from pain and suffering takes into account the context in which such suffering is produced, as we bring forth our expectations and beliefs associated with our conceptions of nature. Oscar Horta (2017) and Catia Faria (2022) have, for instance, called for an urgent reconsideration of the "nature-runs-its-own-course" mindset that justifies non-intervention in the presence of nonhuman animals suffering in the wild. At the same time, the question of intervention has a long and controversial history in photography and filmmaking itself – one that has gradually made its way into discussions about the ethics of wildlife documentary filmmaking. There is still, however, relatively little scholarship about this issue when concerning animal advocacy documentaries, despite the popularity of platforms featuring allegedly genuine videos of rescues and recoveries of (wild) nonhuman animals.[4]

Considering the extent to which much of Kossakovsky's stylistics comply with a feminine aesthetic, it is worth pondering the gender politics of (non-)interventionism, especially when a male director's representation of motherhood is at stake. Faria argues that, in the context of the wild, anthropogenic and naturogenic harms are the result of masculinist idealizations of nature where ecological systems are ethically prioritized over nonhuman animal individuals, who are regarded as "mere elements in the large network of 'wholes'" (2024, 255). "The way in which holistic thinkers lose sight of the individual for the sake of larger, abstract entities," Faria goes on to say, "echoes the traditional, male-based obsession in political philosophy of deriving the individual from society and to accordingly subordinate individual interests for what is thought to be the good of the whole, be it society, the nation or the race" (2024, 257). Perhaps one way to think about the scene of Gunda's killing of the piglet is to ask whether Kossakovsky's double non-interventionism (neither saving the runt nor omitting the scene), which he justifies as an act of respect toward Gunda and her right to be, inadvertently echoes this male-centered form of holistic thinking that trusts nature with a wisdom that should not be questioned or interrupted.

Indeed, in comparing *Gunda*, *Cow* and *The Ghosts in Our Machine*, it is interesting to note how intervention in the face of suffering is represented or implied, and what that suggests regarding the female gaze. The penchant for close-ups in all three films attests to a cinematographic attempt to empathize with their subjects and successfully create a connection with viewers based on intimacy – or, what Derek Bousé (2003) describes as the "false intimacy" resulting from the close-up shot (see Chapter 2). This insistence on intimacy, which is characteristic of the female gaze and the feminine aesthetic, arguably constitutes a form of interventionism itself, in the sense that the filmmakers are responding and "doing something" for their subjects by representing their sentience. And yet, in *Cow*, "[Luma's] death is framed almost flippantly, jammed indecorously into the film's last minutes and shot from a higher angle and a further remove than other scenes" (Schultz-Figueroa 2022, 51). The aesthetics of Luma's death are consistent with Arnold's tiptoeing around an activist agenda and represent the limits of her interventionism as much as a clean break from the intimate type of cinematography that is more readily discerned in the rest of the film. In the case of *The Ghosts in Our Machine*, it is worth noting that perhaps the most poignant words voiced by McArthur, featured just before the title, have to do with the internal conflict of intervention:

> Leaving is always the hardest thing. And leaving is the reason why I am haunted. I'm closing the door as I leave this factory farm, or this fur farm, or this slaughterhouse. I'm closing the door and leaving them behind. Leaving is the hardest part of my work.

It is Marshall's female gaze that symbolically opens the door to the ghosts, creating a means to ensure the permanence of their "looking back" long after they are gone.

Conclusion

The conjoining of Critical Animal and Gender Studies frameworks is helpful for the assessment of how animal advocacy documentaries are devised in terms of subject matter, cinematography, pacing and intimacy with the subjects. This chapter has examined how gendered (and "speciesed") gazes bind the filmmaker's subjectivity with that of their human and nonhuman animal protagonists, and how the *to-be-looked-at-ness* of nonhuman animals opens up an abundance of creative and exegetical choices – choices which demand a consideration of the many levels in which gender politics can come into play. I have argued that documentaries made by male filmmakers tend toward a form of screenwriting where certain formalistic patterns seem to recur, such as the heroic monomyth of the unveiling of secrets that are retrieved for public knowledge, the frequent presence of the filmmaker as protagonist, or the reliance on a narrator to instruct the viewer on polarized positions in the conflict. These elements, fused with a calculated dosage of moral shock prompted by images of violence, persuasively lead toward closed endings in the film narrative (but not to the end of the conflict) that mark the starting point for the viewer to act. The commitment of many male filmmakers to the representation of nonhuman animal victims involves a grand-scale explication of the capitalist forces at work and/or a structural reliance on the principle of quantity (see Chapter 1). Their sense of ethical responsibility toward their subjects implicitly hinders the domination and consumption of the "other" that is characteristic of the male gaze as defined by Mulvey. Documentaries by both male and female filmmakers have envisioned positive models of (eco)masculinities that attempt to break with dominant, hegemonic patterns. Nonetheless, as in the case of *The Game Changers*, variations of the male gaze can filtrate themselves into representations of masculinities that still play by the rules of paradigmatic hegemony and normativity, ultimately circling back to the silencing of nonhuman animals as absent referents.

In contrast with these general patterns, I have analyzed *The Ghosts in Our Machine*, *Cow* and *Gunda* as documentary forms of storytelling that are emblematic of French's female gaze and the feminine aesthetic. Although the monomyth in *The Ghosts in Our Machine* echoes in some ways the heroic journeys that schematize men's films, Marshall's prioritization of relationality, empathy and interconnectedness conveys an ethics of care that is strongly bound to female subjectivity. She accomplishes this by focusing on nonhuman animals gazing back and on McArthur's psychological interiors, for which a strategic representation and framing of spaces becomes necessary. *Gunda* and *Cow* follow suit with their interest in female subjects, mothers in particular. Focalization is determined, for the most part, by their leading figures, Gunda and Luma, as the filmmakers seek to instill feelings of intimacy and empathy by employing diverse tactics affecting camera angles, length of shots, rhythm, soundscapes and montage. The films' endings, far from offering affective closure, suspend viewers in a cathartic sense of loss and nihilism, and represent the culmination of Kossakovsky's and Arnold's respective aesthetic approaches. The comparison of the three films reaffirms French's claim that female gazes (and feminine aesthetics) are plural and multiple. In addition, the analysis has exemplified how animal ethics problematizes gender-based approaches and asks for further study on how gendered aesthetic forms can be effectively used for animal advocacy.

Notes

1 Organizations in the Victorian era (on which American advocates modeled their own) found the bulk of their membership in women, perhaps as a response to growing anxieties in the face of urbanism, the gendered allocation of women in the domestic sphere and the consolidation of vivisection as the standard method of experimental physiology, which contributed to the image of the physician-scientist as a sadistic and unscrupulous male (Turner 1980; Kean 1995; Howell 2000).

2 I have listed *Super Size Me* here because of its critical concern with fast food and the health hazards it poses for humans, which is a subject that is recurrent in other food industry documentaries. Nonetheless, it should be noted that despite the numerous opportunities to seriously delve into the issue of animal ethics and advocacy, Spurlock, for the most part, adheres to an anthropocentric, hegemonic countercultural approach, one that also neglects to consider class and race difference among human consumers (Austin 2007).

3 *Gunda* was also filmed in farms and sanctuaries in Spain and UK. As Kossakovsky explains, the farm that kept Gunda had evolved toward the adoption of welfare-oriented practices and reformed into what he describes as an "empathy farm" (2021b).

4 Popular among these sites is The Dodo, an American media brand and digital publisher founded in 2014 where users document positive encounters with wild and domestic nonhuman animals and upload them as emotionally uplifting and inspiring content.

References

Adams, Carol J. 1990. *The Sexual Politics of Meat. A Feminist-Vegetarian Critical Theory*. Continuum.

Adams, Carol J. 2003. *The Pornography of Meat*. Continuum.

Adams, Carol J., and Josephine Donovan, eds. 1995. *Animals and Women. Feminist Theoretical Explorations*. Duke University Press.

Adewale, Omowale, ed. 2021. *Brotha Vegan: Black Men Speak on Food, Identity, Health, and Society*. Lantern Publishing and Media.

Alaimo, Stacy. 2010. "Eluding Capture: The Science, Culture, and Pleasure of 'Queer' Animals." In *Queer Ecologies: Sex, Nature, Politics, Desire*, edited by Catriona Mortimer-Sandilands and Bruce Erickson, 51–72. Indiana University Press.

Alonso-Recarte, Claudia, and Ignacio Ramos-Gay. 2022. "Real Animals and the Problem of Anthropomorphism in Children's Film." In *The Oxford Handbook of Children's Film*, edited by Noel Brown, 408–28. Oxford University Press.

Andersen, Kip, and Keegan Kuhn, dirs. 2014. *Cowspiracy: The Sustainability Secret*. A. U. M. Films, First Spark Media.

Andersen, Kip, and Keegan Kuhn, dirs. 2017. *What the Health*. A. U. M. Films, First Spark Media.

Antoja, Ed, dir. 2017. *Empatía*. La Diferencia, FAADA.

Arnold, Andrea, dir. 2021a. *Cow*. IFC Films, BBC Film.

Arnold, Andrea. 2021b. "Andrea Arnold on Capturing *Cow*, Bovine Beauty, and the Brutality of Nature. Interview by Rory O'Connor." *The Film Stage*, July 25. https://thefilmstage.com/andrea-arnold-on-capturing-cow-bovine-beauty-and-the-brutality-of-nature.

Arnold, Andrea. 2022. "Andrea Arnold on *Cow*: 'Sometimes When You're Filmmaking, You Get Gifts Sent from the Universe." Interview by Brogan Morris. *BFI*, January 12. https://www.bfi.org.uk/interviews/andrea-arnold-cow.

Austin, Regina. 2007. "*Super Size Me* and the Conundrum of Race/Ethnicity, Gender, and Class for the Contemporary Law-Genre Documentary Filmmaker." *Loyola of Los Angeles Law Review* 40 (2): 687–718. https://ssrn.com/abstract=992217.

Azhar, Danial. 2021. "Who's the 'Wild' One Now? The Exploitation of Non-human Animals as Seen in the Animal Advocacy Documentary *The Ghosts in Our Machine*." *Animal Ethics Review* 1 (1): 19–27.

Bousé, Derek. 2003. "False Intimacy: Close-Ups and Viewer Involvement in Wildlife Films." *Visual Studies* 18 (2): 123–32. https://doi.org/10.1080/14725860310001631994.

Beverly Hills, 90210. Created by Darren Star. Aaron Spelling, prods. 1993–1994. Season 4.

Brinckerhoff, Burt, dir. 1988. "Leaping to Conclusions." *Good Morning, Miss Bliss*. Season 1, episode 4. NBC, December 21.

Campbell, Joseph. (1949) 1968. *The Hero with a Thousand Faces: Bollingen Series*. Princeton University Press.

Carretero González, Margarita. 2023. "Estudios veganos, activismo académico y la paradoja del referente ausente." *Pasajes* 68 (2): 77–88. https://www.jstor.org/stable/27429327.

Carretero González, Margarita. 2025. "Músculos y testosterona a base de plantas: *The Game Changers*." In *Género y sexualidad en el documental animalista y de naturaleza*, edited by Claudia Alonso-Recarte, 31–53. Tirant Humanidades.

Carson, Rachel. (1994) 1962 . *Silent Spring*. Houghton Mifflin.

Clover, Carol J. 1992. *Men, Women, and Chain Saws. Gender in the Modern Horror Film*. Princeton University Press.

Cohen, Mathilde. 2017. "Animal Colonialism: The Case of Milk." *AJIL Unbound* 111: 267–71. https://doi.org/10.1017/aju.2017.66.

Cusack, Carmen M. 2013. "Feminism and Husbandry: Drawing the Fine Line Between Mine and Bovine." *Journal for Critical Animal Studies* 11 (1): 24–45.

Deckha, Maneesha. 2008. "Disturbing Images: PETA and the Feminist Ethics of Animal Advocacy." *Ethics and the Environment* 13 (2): 35–76. https://doi.org/10.2979/ete.2008.13.2.35.

Delforce, Chris, dir. 2014. *Lucent*. Aussie Farms.

Delforce, Chris, dir. 2018. *Dominion*. Aussie Farms.

Derrida, Jacques. 1991. "'Eating Well,' or the Calculation of the Subject. An Interview with Jacques Derrida. Interview by Jean-Luc Nancy, translated by Peter Connor and Avital Ronell." In *Who Comes After the Subject?* edited by Eduardo Cadava, Peter Connor, and Jean-Luc Nancy, 96–119. Routledge.

Derrida, Jacques. (2006) 2008. *The Animal that Therefore I Am*. Edited by Marie-Louise Mallet. Translated by David Wills. Fordham University Press.

Devries, Mark, dir. 2013. *Speciesism: The Movie*. Mark Devries Productions.

DeWaal, Frans. 2022. *Different. Gender through the Eyes of a Primatologist*. W. W. Norton & Company.

Dobbie, Lorena Elke. 2024. "'Please don't look away.' The Ghostly Intersections Between Seeing and Feeling in Documentary Film." In *Animal Activism On and Off Screen*, edited by Claire Parkinson and Lara Herring, 137–54. Sydney University Press.

Donovan, Josephine. 2007. "Animal Rights and Feminist Theory (1990)." In *The Feminist Care Tradition in Animal Ethics. A Reader*, edited by Josephine Donovan and Carol J. Adams, 58–86. Columbia University Press.
Donovan, Josephine. 2011. "Aestheticizing Animal Cruelty." *College Literature* 38 (4): 202–17. https://doi.org/10.1353/lit.2011.0044.
Drew, John. 2016. "Rendering Visible: Animals, Empathy, and Visual Truths in *The Ghosts in Our Machine* and Beyond." *Animal Studies Journal* 5 (2): 202–16.
Duarte, Sofía, and Olimpia Mailat Gurghian. 2025. "Est(ética) de la representación del cuerpo femenino y la maternidad no humana en el cine documental: un análisis contrastive de *Dominion* (2018), *Gunda* (2020) y *Cow* (2021)." In *Género y sexualidad en el documental animalista y de naturaleza*, edited by Claudia Alonso-Recarte, 79–102. Tirant Humanidades.
Dunayer, Joan. 1995. "Sexist Words, Speciesist Roots." In *Animals and Women. Feminist Theoretical Explorations*, edited by Carol J. Adams and Josephine Donovan, 11–31. Duke University Press.
Ehrlich, Pippa, and James Reed, dirs. 2020. *My Octopus Teacher*. Off the Fence, The Sea Project, Netflix.
Faria, Catia. 2022. *Animal Ethics in the Wild. Wild Animal Suffering and Intervention in Nature*. Cambridge University Press.
Faria, Catia. 2024. "Wild Animal Ethics: A Gender-Sensitive Perspective." In *The Routledge Companion to Gender and Animals*, edited by Chlöe Taylor, 255–63. Routledge.
Flynn, Clifton P. 2000. "Why Family Professionals Can No Longer Ignore Violence Toward Animals." *Family Relations* 49 (1): 87–95. https://doi.org/10.1111/j.1741-3729.2000.00087.x.
Frawley, James, dir. 2008. "Life During Wartime." *Grey's Anatomy*. Season 5, episode 6. ABC, October 30, 2008.
French, Lisa. 2022. *The Female Gaze in Documentary Film. An International Perspective*. Palgrave Macmillan.
Gaard, Greta, ed. 1993. *Ecofeminism. Women, Animals, Nature*. Temple University Press.
Gaard, Greta. 2013. "Toward a Feminist Postcolonial Milk Studies." *American Quarterly* 65 (3): 595–618. https://doi.org/10.1353/aq.2013.0040.
Gaard, Greta. 2022. "Toward New EcoMasculinities, EcoGenders, and EcoSexualities." In *Ecofeminism. Feminist Intersections with Other Animals and the Earth*, edited by Carol J. Adams and Lori Gruen, 265–79. Bloomsbury.
Gaarder, Emily. 2011. *Women and the Animal Rights Movement*. Rutgers University Press.
Galkin, Matthew, dir. 2007. *I Am an Animal: The Story of Ingrid Newkirk and PETA*. HBO, Stick Figure Productions.
Gilligan, Carol. 1982. *In a Different Voice. Psychological Theory and Women's Development*. Harvard University Press.
Greenebaum, Jessica, and Brandon Dexter. 2018. "Vegan Men and Hybrid Masculinity." *Journal of Gender Studies* 27 (6): 637–48. https://doi.org/10.1080/09589236.2017.1287064.
Grierson, Tim. 2021. "The Most Riveting Film of 2021 is a Black-and-White Documentary About a Pig." *Insidehook*, April 19. https://www.insidehook.com/film/victor-kossakovsky-interview-gunda-pig-documentary.
Harper, A. Breeze, ed. 2010. *Sistah Vegan. Black Female Vegans Speak on Food, Identity, Health and Society*. Lantern Books.
Hediger, Ryan. 2012. "Timothy Treadwell's Grizzly Love as Freak Show: The Uses of Animals, Science, and Film." *Interdisciplinary Studies in Literature and Environment* 19 (1): 82–100. https://doi.org/10.1093/isle/iss025.
Herzog, Werner, dir. 2005. *Grizzly Man*. Lions Gate Films, Discovery Docs, Real Big Productions.
Hockenhull, Stella. 2017. *British Women Film Directors in the New Millennium*. Palgrave Macmillan.
Horta, Oscar. 2017. "Animal Suffering in Nature. The Case for Intervention." *Environmental Ethics* 39 (3): 261–79. https://doi.org/10.5840/enviroethics201739320.
Howell, Philip. 2000. "Flush and the *Banditti*: Dog-Stealing in Victorian London." In *Animal Spaces, Beastly Places. New Geographies of Human-Animal Relations*, edited by Chris Philo and Chris Wilbert, 35–55. Routledge.

Hunnicutt, Gwen. 2020. *Gender Violence in Ecofeminist Perspective. Intersections of Animal Oppression, Patriarchy and Domination of the Earth.* Routledge.
Innocenzi, Giulia, and Pablo D'Ambrosi, dirs. 2024. *Food for Profit*. Pueblo Unido.
jones, pattrice. 2011. "Fighting Cocks: Ecofeminism Versus Sexualized Violence." In *Sister Species. Women, Animals, and Social Justice*, edited by Lisa Kemmerer, 45–56. University of Illinois Press.
Jones, Rebecca. 2021. "A Recipe for Heganism? Masculinity, Meat-Avoidance and the Cookbook." *Sociální studia/Social Studies* 18 (2): 105–21. https://doi.org/10.5817/SOC2021-2-105.
Kean, Hilda. 1995. "The 'Smooth Cool Men of Science': The Feminist and Socialist Response to Vivisection." *History Workshop Journal* 40 (1): 16–38. https://doi.org/10.1093/hwj/40.1.16.
Kenner, Robert, dir. 2008. *Food, Inc*. Magnolia Pictures, Participant Media, River Road Entertainment.
Kheel, Marti. 2008. *Nature Ethics. An Ecofeminist Perspective.* Rowman & Littlefield.
King, Ynestra. 1995. "The Ecology of Feminism and the Feminism of Ecology." In *Readings in Ecology and Feminist Theology*, edited by Mary Heather MacKinnon and Moni McIntyre, 150–60. Sheed & Ward.
Knight, Julia. 1992. *Women and New German Cinema*. Verso.
Kossakovsky, Victor, dir. 2020. *Gunda*. Sant & Usant, Louverture Films.
Kossakovsky, Victor. 2021a. "*Gunda*: A Conversation with the Filmmaker Victor Kossakovsky." *YouTube* video, 48:29. Posted by *Chicago Humanities*, April 14. https://www.youtube.com/watch?v=0LKY0aOKa8Y.
Kossakovsky, Victor. 2021b. "*Gunda* (doc) – Q&A | Filmmaker Victor Kossakovsky | Film Independent Presents." *YouTube* video, 51:17. Posted by *Film Independent*, February 2. https://www.youtube.com/watch?v=3NWlbTIk0Yg.
Kruse, Corwin R. 1999. "Gender, Views of Nature, and Support for Animal Rights." *Society and Animals* 7 (3): 179–98. https://doi.org/10.1163/156853099X00077.
López Portillo, Yeray, dir. 2018. *Yo Galgo*. Skinny Dog Films, Yo Galgo Productions S.L.U.
Luke, Brian. 2007. *Brutal. Manhood and the Exploitation of Animals*. University of Illinois Press.
Malamud, Randy. 2012. *An Introduction to Animals and Visual Culture*. Palgrave Macmillan.
Marshall, Liz, dir. 2013. *The Ghosts in Our Machine*. Ghosts Media, LizMars Productions.
Marshall, Liz. 2024. "Meat the Future. An Interview with Liz Marshall. Interview by Claire Parkinson." In *Animal Activism On and Off Screen*, edited by Claire Parkinson and Lara Herring, 209–21. Sydney University Press.
Merchant, Carolyn. 1990. *The Death of Nature: Women, Ecology, and the Scientific Revolution.* HarperCollins.
Mills, Brett. 2012. "The Animals Went in Two By Two: Heteronormativity in Television Wildlife Documentaries." *European Journal of Cultural Studies* 16 (1): 100–14. https://doi.org/10.1177/1367549412457477.
Monson, Shaun, dir. 2005. *Earthlings*. Nation Earth.
Morell, Virginia. 1993. "Called 'Trimates,' Three Bold Women Shaped their Field." *Science* 260 (5106): 420–25. https://doi.org/10.1126/science.260.5106.420.
Mota-Rojas, Daniel, Stefany Monsalve, Karina Lezama-García, et al. 2022. "Animal Abuse as an Indicator of Domestic Violence: One Health, One Welfare Approach." *Animals* 12 (8): 977. https://doi.org/10.3390/ani12080977.
Mulvey, Laura. 1975. "Visual Pleasure and Narrative Cinema." *Screen* 16 (3): 6–18. https://doi.org/10.1093/screen/16.3.6.
Newkirk, Ingrid E. 2011. "Are You Waving at Me?" In *Sister Species. Women, Animals, and Social Justice*, edited by Lisa Kemmerer, 65–71. University of Illinois Press.
Nichols, Bill. 1991. *Representing Reality*. Indiana University Press.
Pick, Anat. 2011. *Creaturely Poetics. Animality and Vulnerability in Literature and Film.* Columbia University Press.
Pierce, Kathleen 2010. "Men Leave Their Mark on Veganism." *Boston Globe*, March 24. http://www.boston.com/lifestyle/food/articles/2010/03/24/men_leave_their_own_mark_on_veganism/?page=2.

Plumwood, Val. 1993. *Feminism and the Mastery of Nature*. Routledge.
Porter, Pete. 2023. "*Cow* and *Gunda*." *Society and Animals* 31 (4): 561–65. https://doi.org/10.1163/15685306-bja10143.
Psihoyos, Louie, dir. 2009. *The Cove*. Ocean Preservation Society, Diamond Docs, Skyfish Films.
Psihoyos, Louie, dir. 2015. *Racing Extinction*. Okeanos, Discovery, Oceanic Preservation Society.
Psihoyos, Louie, dir. 2018. *The Game Changers*. ReFuel Productions, Oceanic Preservation Society, Diamond Docs.
Randall, Thomas E. 2018. "Heganism." *Between the Species* 22 (1): 214–35. https://digitalcommons.calpoly.edu/bts/vol22/iss1/6.
Rees, Amanda. 2007. "A Reflection on the Field – Primatology, Popular Science and the Politics of Personhood." *Social Studies of Science* 37 (6): 881–907. https://doi.org/10.1177/0306312707077368.
Reijnen, Ambra. 2023. "Luma and Gunda Stare Back: Integrating Non-Human Perspectives in the Ecofeminist Documentary." *Animal Ethics Review* 3 (1): 1–10. https://doi.org/10.31009/aer.2023.v3.i1.02.
Roscher, Mieke. 2024. "Animal Rights Activism is (Not) a Crime. Portraying Violence and the Animal Rights Agenda in German, American and British TV Crime Series (c. 1990–2010)." In *Animal Activism On and Off Screen*, edited by Claire Parkinson and Lara Herring, 15–37. Sydney University Press.
Roughgarden, Joan. 2004. *Evolution's Rainbow: Diversity, Gender, and Sexuality in Nature and People*. University of California Press.
Sánchez, Soraya F., Cristina Mateos Casado, and Isabel Tajahuerce Ángel. 2022. "Maltrato animal, violencia vicaria y violencia de género: la integración de recursos animalistas en la intervención integral en violencia de género en el ámbito de la pareja o expareja." *International Visual Culture Review / Revista Internacional de Cultura Visual* 12 (1): 1–15. https://doi.org/10.37467/revvisual.v9.3722.
Schultz-Figueroa, Benjamín. 2022. "Death by the Numbers: Factory Farms as Allegories in *Cow* and *Gunda*." *Film Quarterly* 75 (4): 47–57. https://doi.org/10.1525/fq.2022.75.4.47.
Sheehan, Paul. 2008. "Against the Image: Herzog and the Troubling Politics of the Screen Animal." *SubStance* 37(3): 117–36. https://doi.org/10.1353/sub.0.0023.
Spurlock, Morgan, dir. 2004. *Super Size Me*. The Con, Studio On Hudson.
Stein, Jenny, dir. 2000. *The Witness*. Tribe of Heart.
Stein, Jenny, dir. 2009. *Peaceable Kingdom: The Journey Home*. Tribe of Heart.
Stewart, Rob, dir. 2006. *Sharkwater*. SW Productions, Diatribe Pictures, Sharkwater Productions.
Stewart, Rob, dir. 2018. *Sharkwater Extinction*. Sharkwater Productions.
Tabrizi, Ali, dir. 2021. *Seaspiracy*. A. U. M. Films, Disruptive Studios.
Turner, James. 1980. *Reckoning with the Beast. Animals, Pain and Humanity in the Victorian Mind*. Johns Hopkins University Press.
Twine, Richard. 2024. "Ecomasculinities, Boyhoods and Critical Animal Pedagogy." In *The Routledge Companion to Gender and Animals*, edited by Chloë Taylor, 437–48. Routledge.
Velasco Sesma, Angélica. 2017. *La ética animal. ¿Una cuestión feminista*? Ediciones Cátedra.
Vernon, Kathleen M. 2017. "Sound and Gender." In *The Routledge Companion to Cinema and Gender*, edited by Kristin Lené Hole, Dijana Jelača, E. Ann Kaplan, and Patrice Petro, 47–56. Routledge.
Warren, Karen J., ed. 1997. *Ecofeminism. Women, Culture, Nature*. Indiana University Press.
Warren, Karen J. 2000. *Ecofeminist Philosophy*. Rowman & Littlefield.
Watkins, Ed, prod. 2024. *Queer Planet*. Plimsoll Productions.
Wrenn, Corey Lee. 2016. "An Analysis of Diversity in Nonhuman Animal Rights Media." *Journal of Agricultural and Environmental Ethics* 29 (2): 143–65. https://doi.org/10.1007/s10806-015-9593-4.
Wright, Laura. 2015. *The Vegan Studies Project. Food, Animals and Gender in the Age of Terror*. University of Georgia Press.

5 The Instrumentalization of Celebrity Culture

The ubiquitousness of celebrities in twenty-first-century mass media and humanitarian campaigns offers animal advocacy organizations opportunities to assimilate and instrumentalize celebrity culture for their own ends and increase the exposure of their message. This chapter explores the alliance between celebrities and animal advocacy films, and examines the strategies employed in a series of documentaries to maximize benefits deriving from celebrities' visibility and solidarity capital. It deals only with human celebrities, leaving the topic of nonhuman animal celebrities and celebrification as complementary commentary that also interlinks with Chapter 2's analysis of nonhuman animals as biograph-able subjects. The first section provides a critical overview of the celebrity apparatus to better understand the theorizations surrounding their typology and the stakes involved in animal advocacy and activism. Drawing on the field of Celebrity Studies, it unpacks the cultural construction of celebrity as a commodity equipped with different types of capital that affect the social movements they become associated with. The second section discusses how animal advocacy films have strategized about celebrity bodies, faces and voices in their exploration of documentary form and narratology. I address a sample of films to examine the politics of consensualism, that is, the communicative schemes that naturalize celebrity participation in a way that makes the message more palatable for audiences and protects the celebrity brand. In the final section, I approach the figure of Joaquin Phoenix, a critically acclaimed actor that is a recognizable name in animal advocacy films and campaigns. I examine his celebrity as a case where public appreciation of his craft and the authenticity that is part of his brand become intertwined with his activism. His role as narrator in Shaun Monson's canonical film, *Earthlings* (2005), foregrounds the centrality of nonhuman animals as moral subjects. In the film, Phoenix practices forms of consensualism that protect his credibility and standing in the animal rights movement by drawing on American (and Western) ideals of individualism and meritocracy.

The Celebrity Apparatus: Typologies and Stakes

Celebrities have a history of participating in causes that carry a social, political or environmental message. As the "powerful elite" they have become, they are entrusted with the "expertise to perform a moral discourse" that "brings into focus

DOI: 10.4324/9781032661445-6

the inherent theatricality of humanitarianism as an arrangement between those who watch at a distance and those who act on the spot" (Chouliaraki 2013, 78–79). With the diversification of media outlets, scholars have shown a deeper interest in the mechanisms and typologies that articulate celebrification – that is, "the process by which ordinary people or public figures are transformed into celebrities" (Driessens 2012, 643) – and the celebrity apparatus. The exploration of these ecologies has led to the theorization of consumer perceptions of celebrity authenticity, public favorability and likeability.

The successful packaging of celebrity involves forms of representation of personalities where traits such as "star quality" and "charisma" are manufactured into a profitable product that sells. When celebrities lend their faces, bodies and voices to causes, these elements become subjected to ethical evaluations of the core structures of their persuasive endeavors, where audience perceptions of their real motivation, loyalty, consistency, perseverance and transparency will impact their level of credibility – a credibility that is not infallible and is subjected to whims of public opinion. Public monitoring and policing of loyalty to a cause is to a great extent carried out thanks to the widespread use of social media and its exposure of private life (Marwick and Boyd 2011).

Celebrity, therefore, is an unstable category rendered vulnerable by affective and economic capital. In its moments of high favorability, it can be a highly profitable commodity that makes the risk of investment in its fabrication worth it for mass media, entertainment and sports industry professionals, if we take these industries as the leading factories of celebrity tailoring, as suggested by David Meyer and Joshua Gamson (1995, 184). This risk is also one that advocates and activists take when recruiting a celebrity, and so, in a post-postmodern scenario of digital mediatization where causes are packaged and framed into soundbites, social media posts, viral images and tokens of infotainment to maximize the affective response to a fleeting message, the instrumentalization of celebrities becomes a carefully calculated affair. Dan Brockington (2021, 254–55) regards the 2000s in the UK (and perhaps earlier in the United States) as a turning point in celebrity advocacy, when liaison officers began to dynamize the industry in a more organized and professional way, facilitating alliances between celebrities and organizations, charities, NGOs and platforms. This upgrading of clout management has shifted the paradigms of networking in effective ways, where celebrity agencies remain in close contact and negotiation with NGOs and charities, optimizing the matchmaking between celebrity and cause.

The deconstruction of celebrity activism demands an assessment of the types of capital and celebrities at play. The manufacturing of celebrity personalities into consumable icons (or iconic consumables) generates economic capital and symbolic capital, that is, the "prestige, admiration, and authenticity, [that] can translate into economic capital" (Farrell 2020, 6). Authenticity relies on public perception of "consistencies between the celebrity's private and public personas" (Farrell 2020, 6). When activism becomes a visible part of a celebrity's public presence, this consistency will often be measured by perceptions of their commitment to live by what they "preach" – a commitment that is reaffirmed by a timeline dotted with ratifications of their stand on the cause. By spectacularizing lifestyle choices, endorsing certain products and supporting political and public figures or organizations, celebrities and their teams try to control the narrative that construes their persona. They manufacture a biography through careful management of the mediatized forms of iconicity, indexing and framing of the celebrity's means of self-representation. It is precisely such manufacturing (or even the very reminder that celebrities are products and commodities) that invite suspicion on the part of critics as to the celebrity's actual commitment to a cause, or whether their involvement is just a means to strengthen their brand and fashionably parade their ethics in acts of "conspicuous redemption" (Boykoff and Goodman 2009) that are theatricalized in the public sphere.

French sociologist Pierre Bourdieu's (1986) taxonomizing of capital into economic, cultural and social types is amplified in Bruno Campanella's (2020) deconstruction of celebrity. Bourdieu understands social hierarchization as being dependent on both economic and non-economic factors that define and articulate exchanges and relationships. The non-economic assets (cultural and social capital) can, in Bourdieu's view, be converted into economic capital under given circumstances. Cultural capital is the set of assets associated with knowledge, education, skills, taste and lifestyle that can aggrandize the individual's social status and provide them with professional (and therefore economic) opportunities. Social capital, on the other hand, refers to an individual's assets resulting from their network, relation to and membership in given groups. In considering celebrity, Campanella adds to this list Nathalie Heinich's (2012) "visibility capital," which is measured by the exposure and recognizability (and, therefore, ubiquitousness) of the individual, and is also potentially transformable into economic capital. With this in mind, Campanella proposes what he terms "solidarity capital." Solidarity capital is connected to "the moral field," and is "created through the participation of celebrities in media actions based on clearly defined principles of idealism, selflessness and justice" (Campanella 2020, 27). It "conveys an individual's ability to make himself or herself visible in the media along with their capacity to publicly demonstrate exemplary moral behaviour" (2020, 30).

Types of capital intersect with celebrity typologies. Perhaps in a gesture intended to subvert Daniel Boorstin's oft-quoted phrase that a celebrity is "a person who is well-known for his well-knownness" (1992, 57), some scholars have proposed a closer examination of the achievements (or what is considered by the public to merit a notable accomplishment, a perception that is readily connected to visibility and solidarity capital) underlying celebrity status. Chris Rojek distinguishes between ascribed, achieved and attributed celebrity: the first is inherited through bloodline; the second "derives from the perceived accomplishment of the individual in open competition" and goes hand in hand with the possession of "rare talents or skills" (2001, 18). In contrast, attributed celebrity is not bound to a unique skill-set, but "the result of the concentrated representation of an individual as noteworthy or exceptional by cultural intermediaries" (Rojek 2001, 18). What may count as an achievement and as deserved celebrity status may vary for different people, of course, and influence favorability. Furthermore, collective appreciation of achievement will depend on the cultural variables that elevate a personality's traits, how they are exercised and how they mirror that cultural ethos. In the United States (and to a great extent in Western culture at large), a noteworthy accomplishment is generally tied to the mystification of individualism, work ethic, meritocracy and Christian and secularized narratives of perseverance in the face of hardships. As this chapter will later discuss, the perception of celebrities as achieved or attributed affects the credibility of their framing of animal advocacy.

Celebrity advocates can act in a number of ways: as spokespeople, patrons, fundraisers or agents in the governance of an organization (Farrell 2020, 2). As part of an elite whose "notoriety has less to do with what they *do*. . . than with what and who they *are*" (Meyer and Gamson 1995, 184; emphasis in original), the prime asset they bring to the cause they support is visibility (1995, 185). Their hyper-individualized image (and/or voice) can be performatively mobilized as part of an endorsement, a campaign, a protest, an announcement or, in our present case, the moving image that animates nonfiction film. Brockington (2008, 2009) studies how celebrities emerge in the politics of environmentalism and conservation and how they navigate the multilayered discourses nurturing the cultural construct of nature. He distinguishes between celebrities who are already recognizable public figures when they turn to conservation efforts – what Graham Huggan calls "conservationist celebrities" (2013, 3) – and people who become celebrities *because of* their conservationist work. However well-intended they may be, these "celebrity conservationists" (many of whom are wildlife and nature documentary presenters, such as Steve Irwin, Sir David Attenborough or Jacques Cousteau) are part of the corporate, capitalist apparatus that has created them and that also packages "nature" as a consumable (Brockington 2009). The commodity of "closeness to nature" (Brockington 2008, 561) is manufactured with the instrumentalization of charismatic celebrities that compensate audiences' sense of alienation from these spaces (2008, 562).

Building on Brockington, studies on the connections between celebrity and conservation have proliferated in the last couple of decades due to the climate crisis, the greening of companies and corporations and the diversification and democratization of celebrity. If anything, the variety of themes and scopes of these "celebrity ecologies" (Goodman and Littler 2013) demonstrates the efforts invested in the celebrification of nature and nonhuman species and in the branding of the celebrity as a conservationist. In order to understand these processes, scholars attempt to obtain data and quantify how the alliance with celebrities can benefit, damage or be inconsequential to an environmental cause. For example, Katie Blake et al. (2023) have studied how species eponyms based on celebrity names can impact the visibility and popularity of certain taxonomic groups, and echo how in the twenty-first century species etymologies have been influenced by pop culture (Mammola et al. 2023). Other studies discern and probe the limitations of measuring the effectiveness of celebrity endorsement in environmentalist causes (Olmedo et al. 2020) and offer a working guide to better inform decision-making around celebrity recruitment (Olmedo et al. 2021). Maxwell Boykoff and Michael Goodman (2009) explore celebrities' appropriation of climate change expertise against a backdrop of critical skepticism regarding their authority and effectiveness, while Elizabeth Duthie et al. (2017) note the distracting effect of a celebrity endorser. Sian Sullivan focuses on how celebrity involvement contributes to the "spectacle" of conservation that elevates pristine nature at the expense of the actual people that inhabit these landscapes, who are commodified "on terms guided by paying customers" (2011, 341). Patrick Murphy (2021) pries apart celebrity conservationist taxonomies to make room for the type of youth-centric ecopolitics embodied by Greta Thunberg, while Theresa Winge's (2008) historicization of eco-fashion cautions against the way celebrities have depoliticized many of the controversies surrounding the exploitation and resource management of the fashion industry. With a particular focus on American movie star and standard-bearer Leonardo DiCaprio, Richard Maxwell and Toby Miller (2013) and Jackie Raphael (2020) denounce the inconsistencies between the very public conservationism of celebrities and the detrimental environmental effects caused by their private lifestyles and the on-location shooting of the films they star in. Similarly, and echoing Maxwell and Miller (2012), Manel Jiménez-Morales and Marta Lopera-Mármol (2022) examine green shooting and the need for the media and entertainment industries to adopt sustainable practices beyond the smokescreen of greenwashing.

Celebrity alignment with conservation previews some of the issues that arise in vegan and animal activism. But because vegans, more so than conservationists, are often negatively perceived as deviants and preachy moralists, if not radicals and extremists (Cole and Morgan 2011; De Groeve et al. 2021; Brookes and Chałupnik 2022; Rutledge-Prior 2023; Napoli and Ouschan 2024), celebrity associations with veganism are, it is my contention, subjected to a higher risk and vulnerability that can threaten economic, affective and social capital. In the 1990s, Meyer and Gamson listed the concern for a clean environment as an example of a less confrontational type of cause, where anyone, including celebrities, "have an obvious stake, or standing, in the outcome of such a political dispute, and therefore can make claims on government" (1995, 190). Evidently, environmentalism can also stretch into far more radical demands, and the movement has since evolved toward media representations of ecological apocalypticism. But the perception of vegans is generally associated with more extreme positive or negative ends of the spectrum, being conceived as either respectable practitioners of moral beliefs or, perhaps more frequently, arrogant and intransigent fanatics (De Groeve et al. 2022). This results in vegans' perception of themselves as targets of discrimination (Reuber and Muschalla 2022). Such discrimination can ultimately affect the effectiveness of the movement, with evidence suggesting that carnists are more likely to be persuaded to reduce their consumption of meat when the messenger is an omnivore than when the messenger is vegan (Thürmer et al. 2022).

When a social movement presents such risk to the capital and reputation of celebrities, leveraging fame and status is often reframed into more consensual claims that can allow them to maintain their standing, that is, their legitimacy to become part of that public conversation (Meyer and Gamson 1995). Framing a cause more consensually might involve remodeling the movement's claims into charity, where "the claimant calls for generosity, but not for any kind of structural political change" (1995, 190). Another strategy is having celebrities subsume the claims "into larger collective ones" (1995, 190). Ethical veganism (i.e., refraining not only from a diet of meat, eggs and dairy but also from the consumption of products tested on nonhuman animals, fabricated with by-products, or involving the needless and noxious spectacularization of their bodies) is often made more consensual by reframing it into discourses on healthy plant-based diets or diluting it in environmentalism, at the expense of the objectification and neglect of nonhuman animals as moral subjects (Spang 2024). As of the late 2010s, this discursive softening has greatly facilitated the emergence of veganism in mainstream Western culture, with an urban growth in the number of vegan restaurants and services, along with vegan products and brands that make it fashionable and desirable (Nguyen 2017) – and celebrities are both instigators and practitioners of the trend. Celebrities can legitimize their standing by framing their veganism as a lifestyle choice and propagating uplifting discourses about self-care that are disseminated in social media (Doyle 2016; Phua et al. 2020). The entrepreneurship and capitalist vision that promote lifestyle advocacy, however, threatens to reproduce the same neoliberal, consumerist, corporate logic that ethical veganism, as a philosophy, seeks to dismantle – what Sexton et al. (2022) term "Big Veganism" in the context of food geographies.

The very category of celebrity, of course, plays into these contradictions as well. As signifiers of postcapitalist consumer culture, their status is not built on institutional power (Farrell 2020, 4), but on power and influence deriving from their profitability as marketable and tradeable commodities meant to sell and amass wealth. How these tensions affecting authority, image, branding and the capital at stake are handled in documentary film is revealing of the extent to which filmmakers may or may not comply with mechanisms of the consensual, hereafter referred to as "consensualism."

Celebrity Narrators, Faces and Bodies

Animal advocacy documentaries often position celebrities as the face and/or voice of the movement, but the visibility they confer should be critically evaluated. To begin with, celebrities, as the intertexts they are, carry their reputational "baggage" into the advocacy projects, and so, insiders within the movement, may perceive this "migration" into vegan campaigning as potentially detrimental if the celebrity in question lacks broad public favorability (Cherry 2024). It must also be remembered that visibility is neither an end in itself nor does it guarantee the successful completion of the movement's message (Olmedo et al. 2020), which is to motivate viewers toward actual (and ideally, immediate), effective change. There is certainly a likelihood that the more exposure a message gets (which is what the celebrity mostly supplies), the higher its chances of effectiveness; yet measuring documentary effectiveness is a complex task in need of reassessment of methodologies and tool availability (Karlin and Johnson 2011).

In contrast to the measurement of success of commercial fictional films, which is commonly reduced to box office profit relative to budget – and which is perhaps more indicative of successful marketing and promotion (Whiteman 2004) – activist and advocacy documentaries are defined as subgenres whose goals are to educate and mobilize audiences. This is difficult to fully measure (the "*Blackfish* effect," discussed in Chapter 2, is an exceptional case, not the norm). Most methodological approximations are reliant on the results of data obtained in pre- and post-screening interviews, surveys and questionnaires. Furthermore, open-ended questions can lead to suggestive but indefinite answers on the part of participants (Bieniek-Tobasco et al. 2019, 13), where initiative and long-term commitment are hard to confirm.

With this in mind, the questions posed here are how celebrities manage their image and leverage their capital with the knowledge that their support increases the visibility and exposure of the documentary, and how, at the same time, they navigate the politics of the consensual to protect both their standing and their brand. Under the auspice of filmmakers and with change and mobilization as the objectives, celebrities lending their names, faces and/or voices as synecdoches of their persona to animal advocacy documentaries manage consensualism through different framings. For example, Louie Psihoyos's *The Game Changers* (2018), discussed in Chapter 4, dilutes arguments of animal ethics in favor of those promoting health (and hegan) culture. The constellation of champion athletes in the cast, along with the list of entertainment and sports industry titans featured as executive producers (director James Cameron, actors Arnold Schwarzenegger and Jackie Chan, and sports icons Lewis Hamilton, Novak Djokovic and Chris Paul) create a choral structure of influential celebrity capital. Whatever individual takes on animal ethics arguments these celebrities might have, these are dissolved in the film to cohesively present a straightforward, unified voice that speaks to the interests of the public in leading a healthier lifestyle. As far as vegan documentaries go, the film's type of consensualism problematizes its categorization as a production committed to ethical veganism.

Consensualism can also be approached through form, by tapping into viewers' expectations of how content should be presented and ordered. The classic biographical genre, in its most conservative form, is one such example, as it offers conventions that satisfy viewers' anticipations. When portraying a celebrity, filmmakers can play with the "relatable yet unreachable" paradox they elicit by devising structures that respond to the shared expectations of viewers as to how life stories should be told. They give the audience a form they feel they have control over, easing the digestion of content. Representations of the singled-out individual as a historical agent will usually resort to narrative arcs marked by stages of initiation, journey (trials and tribulations) and return that settle viewers in the comfort zone of recognizable anthropocentric monomyths (Campbell [1949] 1968). Activism and advocacy, which are acts that can only take place because there is disagreement, confrontation or lack of consensus preceding them, offer an ideal thematic axis from which to mold such narrative arcs. Rachel Kahn and François Chaumont's television documentary *Brigitte Bardot, Le Serment fait aux Animaux* (2019), for example, is a standard representation of Bardot's life and achievements with her drastic shift into animal rights activism as centerfold. Audience familiarity with traditional biographical form involving the expository mode, use of archival footage and interviews makes Bardot's animal rights message more palatable, and the uplifting images of rescued animals and the comparatively smaller number of images of animal cruelty are made meaningful as they are deployed for the representation of her persona. Hypothetically, this can have a more positive effect (viewers are inspired to emulate Bardot), or a more negative one, as the glamorization of the celebrity subject by sheer virtue of the film's structural attention not only depoliticizes institutionalized speciesism but also inhibits viewers from introspectively surveying their own complicit passiveness toward animal exploitation. Another example is Matthew Galkin's depiction of Ingrid Newkirk in *I Am an Animal: The Story of Ingrid Newkirk and PETA* (2007). But although more politically invested in its framing (perhaps because the subject is a celebrity activist, and not an activist celebrity, if we are to draw an equivalence with Brockington's typology), the film still historicizes its human protagonist in a manner that downplays the "biograph-ability" (see Chapter 2) of nonhuman animal subjects.

Attempts at consensualism through conventional, anthropocentric biographical form and content, therefore, cannot be carried out without considerable sacrifices. What the biographical genre shows is that celebrities (whose popularity justifies the interest in creating a biography in the first place) can easily overshadow or mitigate a movement's message, in the same way that celebrity endorsement can backfire by reducing recall and attention to the brand or product (Pelau et al. 2022; D'Ambrogio et al. 2022, 724), leading to what in advertising is known as the "vampire effect" (Erfgen et al. 2015). Of course, one could argue that it was never the filmmakers' intention to make an advocacy film, and that the focus was always meant to be the celebrity's life, which challenges assumptions about strategic consensualism and the films' categorization as activist documentaries. There are indeed examples of documentaries that engage with history and celebrities where the filmmakers' intentionality is clearly an activist one, such as *Punk Rock Vegan Movie* (2023), written and directed by Moby, which offers a communal biography of straight edge and punk rock musicians. Moby's historicizing of ethical veganism in the punk rock movement is also carried out through conventional techniques such as interviews and archival footage, and he lightens his own authoritative tone with comical scenes in which he converses with his dog, Bagel. In this way, the historiography he offers appeals to audiences by creating a familiar, amicable mood – a consensual gesture itself. Conventional forms of the biographical and historical genres, therefore, present celebrity activists and activist celebrities as sources of inspiration, as viewers may be moved to follow in the footsteps of their "heroic" role models. From an activist perspective, however, these forms can be perceived as egoistic and self-gratifying, compromising the centrality of nonhuman animals as moral subjects.

The biographical and historical genres are not, in any event, a prerequisite to maximize the affective capital of celebrities in animal advocacy films. Many documentaries naturalize the celebrity's moral stand by decontextualizing it from a personal, chronologically ordered life story. Rather than narrativizing how the celebrity came to believe and practice ethical veganism, they represent them as if they have always and will always follow their moral code. The celebrity is meant to appear as a permanent, unbreakable force – an effigy of credibility. In these instances, the celebrity presence is enacted through vocal and/or facial performance as they take on the role of narrators and presenters (rather than appearing as interviewees). Quite often, celebrity narrators appear in documentaries intended to inform and affectively engage the viewer by causing moral shock (see Chapter 1). The use of present verb tenses in the expository mode is significant here: narrators' emphatic insistence that nonhuman animals *are* suffering (as opposed to the biographical genre's chronological focus on the past) strengthens the sense of urgency and projects the moving image as an index of ongoing practice, dehistoricizing it from the moment when it was recorded and coherently eternalizing it in the perpetuity of the animal-industrial complex. Narrators employ slow, grave tones that strategically place silences in moments that grant the images full protagonism. The sobering qualities of narrators' vocal performance are energized by affective capital: viewers' familiarity with the celebrity translates into an intimacy of how they *presently feel*. It is as if they were watching and experiencing the image at the same time as the viewer, inferring that affinity with the celebrity will depend on being moved to a similar affective state. Feature-length and short films in this line are common in animal advocacy (stylistic variations notwithstanding), with celebrity narrators such as Julie Christie in *The Animals Film* (Schonfeld and Alaux 1981), Joaquin Phoenix in *Earthlings* (Monson 2005), James Cromwell in Mercy for Animals' *Farm to Fridge* (Iovino 2011) and Natalie Portman in *Eating Animals* (Quinn 2017) serving as prominent examples. Sometimes filmmakers' resort to a thread of celebrity narrators, rhythmically intervening in different sections and muscling their authority by sheer number. Chris Delforce's *Dominion* (2018) features popular entertainment industry celebrities Rooney Mara, Sia, Kat Von D and Sadie Sink complementing Joaquin Phoenix as narrators, and Shaun Monson's film following *Earthlings*, significantly titled *Unity* (2015a), features a star-studded string of 100 narrators.

In most animal advocacy documentaries with (celebrity) narrators, the reaffirmation of the expository mode is also achieved by avoiding traditional anthropomorphist mechanisms. Narrators rarely communicate as if they were the nonhuman animals in the images; they do not speak from the position of the nonhuman "I." There may occasionally appear a literary citation with a nonhuman animal speaker, but these are meant as performative beautifiers. When nonhuman animals do appear heavily drenched in anthropomorphism, this is usually intended as a comic stylization of the filmic tone – an example of this is Bagel, Moby's dialogic canine partner in *Punk Rock Vegan Movie.* For the tone to remain recognizably somber, therefore, linguistic anthropomorphism is generally avoided. In contrast, in environmental films such a strategy is not unheard of: Ulrike Olausson and Ylva Uggla (2021) analyze Conservation International's 2014 "Nature Is Speaking" campaign, made up of twelve films narrated by Hollywood A-listers that celebrify nature by speaking as if they were the vocal embodiments of her elements (Mountain, Ice, The Rainforest, etc.). When celebrification operates in the nonhuman context, an anthropomorphic projection of human traits onto nonhuman entities takes place

(Olausson and Uggla 2021, 354–55; Giles 2013). The media's power to engineer nonhuman celebrity involves the attribution of human qualities to create a personality for the public to engage with and consume (Giles 2013). The fact that animal advocacy documentaries do not generally employ narrators that speak as if they were nonhuman animal subjects (but do speak on their behalf as third persons) illustrates the movement's overall suspicions about the effectiveness of anthropomorphism, as these projections are mostly associated with disneyfied and infantile representations that can attenuate an ethical response on the part of viewers. Were a celebrity voice added to the anthropomorphizing formula, the risk of a celebrified caricature would be too great. Celebrities are not empty signifiers; they are experienced by the public as intertexts that via proximity can infuse other people, beings or objects with traits associated with their personality and physique. To formally distance celebrity from nonhuman animals just enough to ensure the visibility of the documentary seems a plausible reason as to why filmmakers resist approaching the narrative voice in ways that could potentially make a humanoid celebrity out of their subjects and violate their integrity.

When celebrity narrators do not physically appear on camera, they transfer their affective (and, paradoxically, their visibility) capital onto soundscapes. In this manner, the potential visual distraction of the celebrity body is prevented and the voice-of-God authority of the expository mode is consolidated. The films promote empathetic connections with the embodied nonhuman animals that occupy the moving image without the reminder of the physical and biological similarity between viewer and celebrity at a corporeal level. Corporeal identification can be carried out with human "characters" that are polarized into benevolent and malevolent agents: viewers can identify with rescuers or activists while feeling repelled by the species identity they share with exploiters. Human bodies are very important for the consummation of moral shock, because they foster strong reactions of (dis)identification with morally categorized actions. But the celebrity body is too hyper-individualized to channel this symbolic transference without causing interruptions. As a commodified object of ocularcentrism, it threatens to compete with the visibility of the nonhuman animals with whom it shares the shot.

Making it a point not to compromise the representation of nonhuman animals as moral subjects does not mean that consensualism is forsaken. It means that public approval of the activist celebrity is sought in alternative ways that can still preserve the ethical vegan agenda. One such way is by associating ethical veganism with welfarist approaches (perceived as less threatening by the general public) and forming a common front. In Christopher Quinn's *Eating Animals* (2017), based on Jonathan Safran Foer's 2009 *New York Times* bestseller, ethical vegan arguments are placed alongside (but not subsumed into) the views of farmers and whistleblowers whose anti-corporativism is nostalgically rooted in a romantic, preindustrial past when workers "loved" and felt responsibly (and aesthetically) connected to the nonhuman animals they raised for food. Resistance to tyrannic corporativism and the establishment can be read in the key of Americanness, as discussed in Chapter 3. With a common enemy in sight, viewers find in Natalie Portman, the narrator, a conciliatory figure that speaks for the spectrum stretching from welfarism to animal rights. As an American diplomat of sorts, her celebrity emerges as an agent of consensualism.

Celebrity narrators can also be deployed as moral authorities when they become grouped as part of a single voice. In a more intensified display of interspecies connection that began with *Earthlings*, Shaun Monson's *Unity* (2015a) ambitiously uses its chorus of celebrity narrators to appeal to the dissolution of the cultural constructs subjecting living creatures and nature to oppression and domination. Names such as Joaquin Phoenix, Anjelica Huston, Amanda Seyfried, Ben Kingsley, Jessica Chastain, Pamela Anderson, Tony Hawk, Russell Simmons, Susan Sarandon, Dr. Dre and Jennifer Aniston are part of the long inventory of actors, television personalities, musicians, writers, athletes, artists and spiritual leaders contributing to the narrative voice. The film drifts, collage-like, along moving and still images of nonhuman animals, humans, landscapes, works of art, historical archives, etc. One minute the film sublimates the idea of a shared consciousness, and the next it shows people at their worst and most destructive. With similar speed, celebrity voices consecutively intervene. Interestingly, images of celebrities are occasionally used to boost the script's more negative comments: a panoramic shot of an Oscars' red carpet ceremony meets the narrator's remark on humans' preoccupation with the "externalization of life," "appearances" and "the material and the individual"; controversial rapper Kanye West is indexed as an example of how humans are obscured by "[possession] of the ego"; extravagant pop star Lady Gaga illustrates the egocentric "mask or appearance one projects to the world"; and another red carpet event is paired with the narrator's question about the meaning of life: "Was it all in pursuit of fame? Killing ourselves for recognition." Significantly, images of celebrities plucked out of representative spaces of Hollywood vanity and exhibitionism are combined with more positive comments: a newlywed Paul McCartney epitomizes the elatedness of "being psychologically nourished by another"; and Angelina Jolie in her humanitarian facet is visually listed as a fulfilled being who has removed "the mask that society puts upon you." As a visual follow-up to Mahatma Gandhi, the Dalai Lama and Nelson Mandela, there is little doubt that Jolie is being cherry-picked as a philanthropist, setting her apart from her celebrity peers, despite the criticism she has often received against her brand of activism. *Unity*'s understanding of celebrity as commodity, therefore, is twofold: the repertoire of celebrity voices means a wealth of visibility capital, and at the same time, celebrity bodies (none of which belong to any of the narrators, as far as I have been able to investigate) effortlessly supply connotations and help the script move forward.

In an interview, Monson revealed that he had removed some footage of animal suffering from the original cut to deliver a more uplifting film, and described the chorus of narrators as a symphony: "There is nothing wrong with a single note"; he explains, "it can be beautiful. But it's not a symphony. And I started thinking of narration as music, as a symphony of voices, and the bigger the chorus the more beautiful the symphony" (Monson 2015b). This symphony, one could additionally argue, also presents a means for celebrities to approach the consensual, for by becoming a hundredth part of the whole, their individuality is subsumed into a higher vocal body that at once speaks about and enacts harmony and agreement. There is little risk here for the celebrity brand, little to cause public discordance with their individual participation. Yet the chorus, symphonic as it may be, can also, with its shifts between narrators, serve as a regular reminder of the scripted nature of the words spoken. This can bring the attention of viewers to the meta-structures of the film (some might even attempt to guess which celebrity narrator is speaking – another instance of the "vampire effect"). While a single or very limited number of narrators promotes, as discussed earlier, a form of intimacy between the self and the celebrity, an ensemble of narrators proposes a connection between the self and the packaged idea of celebrity. From the start of the film, Monson conditions viewers' consumption of the voice-overs. The opening acknowledgment thanks "narrators, musicians, artists and technicians who voluntarily contributed to the overall message of this film, even if some segments did not always reflect their own views." This is followed by the disclaimer that "[t]he conclusions expressed herein are entirely that of the filmmaker." Inadvertently or not, what Monson asks of the viewer is to behold and consume the narrators *aesthetically*, and to put aside the parameters with which we typically survey celebrities as moral figures (authenticity, credibility, legitimacy, etc.). Although not a common strategy in advocacy or activist documentaries, where the audience expects an explicit alignment between the views of the filmmaker and the celebrity, this disclaimer – together with the symphony of voices – precipitates a form of consensualism because celebrity standing and stakes are circumvented, if not pre-excused, redeeming them from any accountability.

In other cases, celebrity narrators do appear on camera, taking on the role of presenters. This strategy is commonly used by People for the Ethical Treatment of Animals (PETA) in their short videos. PETA, of course, has a history of fixation with celebrity bodies and faces, especially since their tactical shift in the 1980s toward enlisting pop culture icons in their campaigns (Simonson 2001). It is also an organization that provides a "safer" platform for celebrities because it has itself become celebrified. Hallmarked in controversy, PETA has garnered considerable Hollywood favorability by circulating the "hottest" stars in their ongoing campaigns. In other words, celebrities can safely strengthen their solidarity capital when following in the footsteps of A-listers who have already lent their image, making it fashionable to participate in that same organization and adding advocacy to their portfolio. The popularity and ease with which PETA navigates the culture industry and mass media as just another player in the game reduces the risk that celebrities face. PETA merits successfully commercializing celebrity advocacy – not a negligible contribution considering that when celebrity messages are perceived as too political, public interest wanes (Miller 2024, 274–75). In their short films, videos and clips, PETA maximizes the star power of the celebrity face. In *Glass Walls* (2009), discussed in Chapter 1, and in *Face Your Food* (2014), Paul McCartney and *Game of Thrones* star Peter Dinklage respectively appear at the beginning and the end of the videos, facing the camera in a medium close-up shot and speaking directly to the viewer. Voice and physiognomy command authority while seeking to inspire compassion and an ethical response to the images of suffering nonhuman animals that make the bulk of the videos. The presenter's addressed gaze, implying eye contact with the viewer, rhetorically echoes the titles of the videos, signifying transparency and direct engagement, while simultaneously strengthening the connection between self and celebrity (Ilicic and Brennan 2020). Celebrity or not, human faces in advocacy documentaries offer viewers an opportunity to exercise their detection of the affective correspondence between what is said and what is felt (see Chapter 3). In the nuances of physiognomy, the viewer traces signs of authenticity and accountability. But the stakes dependent on the celebrity face are in a way higher, for both the social movement and the celebrity. If the celebrity is perceived as inauthentic, or their popularity and likability are at a low ebb, negative responses can rub off on the organization behind the campaign. As to the celebrities themselves, much (if not most) of their livelihood depends on maintaining credibility and favorability, which in turn legitimizes their standing in the movements they publicly endorse.

Celebrity faces and bodies can also become problematic when they contribute to campaigns and films that target foreign traditions. Western celebrities commonly appear in philanthropic documentaries where "views on global relationships [are represented] as a world of 'us' and 'them,' 'the West' and 'the Rest'" (Van den Bulck et al. 2015, 110). When documentaries undertake this angle, they tend to downplay "the historical role of western development in creating problems for what are now developing countries," having "the effect of presenting international aid and philanthropy in benevolent terms as a 'gift'" (2015, 119). Unsurprisingly, many viewers will note the racist objectification of foreign cultures. However, the dichotomy of "us" and "them" can also be treated as part of the rhetoric of consensualism insofar as it again reproduces the familiar myth of the great white savior – a comforting motif for a viewer uncritical of Western ideology and geopolitics.

In Chapter 3 I discussed Louie Psihoyos's *The Cove* (2009) as a film that attracted criticism for its perceived patronization of Japanese customs and the developing countries that supported their interests. The film included footage of an earlier group of activists confronting Taiji fishermen – among them, American actress Hayden Panettiere. A similar "cameo" is made by actress, sex symbol and long-time committed activist Pamela Anderson in Vincent Kelner's *A Taste of Whale* (2022), a documentary that explores both sides of the conflict surrounding the grindadráp (the seasonal mass slaughter of pilot whales and other small cetaceans) in the Faroe Islands. Teaming up with the organization Sea Shepherd France, a heavily made-up, high-heeled Anderson is confronted by journalists who point out her lack of knowledge about how the whales are hunted. Profiling the dialogic opponent as ignorant is a common strategy used by defenders of blood sports and traditions. But when the opponent is a celebrity who struggles (as many other animal advocates would) to prove that they are versed and knowledgeable about the technical and methodological specifics of the tradition, the debate is packaged and exposed as the publicity stunt that it is, hindering the political message. In addition, as Elaine Jeffreys (2016) suggests, another likely reason why campaigns with celebrities fail or are not well received by the targeted culture is that they tend to be more concerned with reproducing international communication strategies than with engaging with local forms of framing. Indeed, the figure of the attractive, blonde actress (say, Panettiere, Anderson or Bardot) protesting and campaigning against foreign local and regional practices of animal exploitation (including Western ones) has brought considerable visibility to the movement, but it has also, in the view of many, reframed colonial discourses into a form of condescending maternalism (Frenette et al. 2023). These glamorized, gendered embodiments are viewed suspiciously by communities that have become increasingly aware of the global presence of Hollywood as a metonymy of American imperialism and hegemony.

Joaquin Phoenix: A Case Study in Authenticity and Consensualism

Among the Hollywood names of ethical vegans that stand out in popular culture (Alicia Silverstone, Natalie Portman, Paul McCartney, Ellen DeGeneres, James Cromwell, Moby, Pamela Anderson, Woody Harrelson, Rooney Mara, Tony Kanal and Ricky Gervais, among others), Joaquin Phoenix is a staple of the movement. Phoenix has made it a point to use public platforms (except for social media accounts, which as of 2025 he continues to avoid) to advocate for nonhuman animals and has regularly participated in protests and campaigns. His long-term commitment, still to this day, greatly reinforces his credibility. A vegan since childhood, he became publicly active in animal advocacy in the 2000s. While many celebrities come and go as fast as campaigns wear out, Phoenix's persistence, overall coherence and consistency have strengthened his legitimacy and standing as an activist. Contrary to the flashy images and soundbites of "conspicuous redemption" (Boykoff and Goodman 2009) that engineers of celebrity circulate in the media to build up the solidarity capital of the personality-brand, Phoenix's persevering presence and comportment in public platforms have earned him a certain right to be taken seriously as an authentic advocate.

It is important here to differentiate between positive perceptions of authenticity and positive responses to the message. The credibility of a celebrity as an authentic advocate may be positive (they are perceived by the public as individuals who genuinely believe in what they say and ethically stand and live by it); but this does not necessarily mean that the public agrees with the celebrity's message or will follow their lead, or that the celebrity is a likeable figure as a professional, an advocate or a personality. Perceptions of the celebrity's authenticity certainly affect audiences' attitudes toward them (Jain et al. 2024; Pöyry et al. 2019), and it is plausible to infer that the more authentic a celebrity is perceived, the more likely it is that the consumer will engage with their ethical message (Quesada Baena et al. 2025). Also, when celebrities are associated with an altruistic endorsement, they tend to be more favorably perceived by the public, and "consumers are more likely to be strongly influenced if they believe that the endorser truly likes or believes in the product" (Kelly et al. 2014, 59). But the correspondence is neither always the case nor systematic in its effectiveness. As discussed earlier, celebrities bring visibility and assist in placing the message front and center as best as they can, but the success of subsequent consumer/viewer engagement with their message is difficult to measure.

While positive perceptions of Phoenix's authenticity and consistency, therefore, may provide an advantageous starting point for the message, they do not in themselves guarantee audience engagement with the cause. In addition, consensualism must still be complied with to be as effective as possible while protecting legitimacy. Phoenix's public advocacy, of which his narration of *Earthlings* is arguably one of his finest accomplishments, shows how celebrities can achieve consensualism without compromising the place of nonhuman animals at the heart of the moral debate. It also shows the nature of celebrity as a cultural construct with a survivalist capacity to adjust as a discourse.

Born in Puerto Rico in 1974, Phoenix has an established reputation as a talented actor with an extensive list of prominent accolades and nominations to his name. These merits justify his categorization as an achieved celebrity (Rojek 2001). His narration of *Earthlings* adds to the film's aura as *the* landmark documentary that sets the bar for all twenty-first-century films to follow. In Chapter 1, I discussed the film's moral shock tactics, and it is important to remind the reader here that *Earthlings* is often credited by advocates – including influential television host (and currently cancelled) Ellen DeGeneres (Doyle 2016, 786) – as one of the impactful reasons why they transitioned into veganism (Fernández 2021; Jones 2022). Phoenix's collaboration with organizations such as In Defense of Animals, Animal Equality, Farm Sanctuary, Los Angeles Animal Save and PETA (who named him Person of the Year in 2019), has garnered him a solid reputation as an animal advocate, putting his visibility and solidarity capital to use in campaigns, protests and nonfiction films. Aside from *Earthlings* and joining fellow celebrity narrators in *Dominion* and *Unity*, he was also an executive producer for *The Animal People* (Suchan and Hennelly 2019), *Gunda* (Kossakovsky 2020) and *The End of Medicine* (Lockwood 2022). The visibility he brings to the animal rights movement cannot be overstated, although there are frictions in his career that have raised eyebrows now and then. One such example is the fact that he rode horses in the western film *The Sisters Brothers* (Audiard 2018), an act he later said he felt bad about (Phoenix 2018). Online responses to the news ranged from disappointment (Phoenix is financially well-off; there was no need to take on the role if it involved horseback riding, which is generally regarded as antithetical to ethical veganism) to sympathy (vegans do their best with what they are given professionally, and it is understandable if he had to be flexible to preserve the role). The first response regards the celebrity as an agent capable of choice and with leverage in the industry; the second, as an individual with limited power that must compromise his belief system in certain circumstances. By the time Phoenix was riding horses again in Ridley Scott's *Napoleon* (2023), some of the criticism was deflected by publicity emphasizing how the actor had a customized plant-based hat and attire made for the film.

Phoenix's acceptance speech at the 92nd Academy Awards in February 2020 for his lead male performance as the titular character in *Joker* (Phillips 2019) concluded an award season in which the actor had used his onstage platform as winner to denounce social and environmental issues and used media scrutiny to draw attention to animal rights (Parkinson and Herring 2023, 585; 588). In all likelihood aware of the buzz created around his probable win, he used his Oscars speech to invoke ethical veganism, and despite the historically low ratings of the ceremony (Katz 2020), the 23.6 million viewers that did tune in meant high exposure for the animal rights movement. Phoenix's reference to the "gift" of being able to "to use our voice for the voiceless" soon landed on the motif of intersectionality, loudly echoing his narration of *Earthlings* and *Unity*: "I see commonality. I think, whether we're talking about gender inequality or racism or queer rights or indigenous rights or animal rights, we're talking about the fight against injustice" ("Joaquin Phoenix Wins" 2020). The thematic angle of the speech went beyond vegetarianism to deliver a powerful visual for ethical veganism at the intersection of female exploitation issues (Parkinson and Herring 2023, 592):

> We feel entitled to artificially inseminate a cow and steal her baby, even though her cries of anguish are unmistakable. Then we take her milk that's intended for her calf and we put it in our coffee and our cereal.
>
> ("Joaquin Phoenix Wins" 2020)

In the last part of his speech, Phoenix came full circle with an uplifting message calling for compassion and growth toward transformative change. Placing himself at the center of a narrative of redemptive masculinity (Parkinson and Herring 2023), he pictured a society of forgiveness and support, and closed his vision by quoting his brother, actor River Phoenix, who tragically overdosed in 1993: "Run to the rescue with love and peace will follow" ("Joaquin Phoenix Wins" 2020).

Analyses of the media reception of Phoenix's speech reveal ambivalent responses. Discomfort with the analogies between gender and race (Parkinson and Herring 2023) met with politically interested coverage that either made a passing note of the vegan message and absorbed it within wider observations on protest speech, dismissed it as ludicrous, or felt it insulted and wronged the livelihood of those in the animal agriculture industry (Loy 2024). Notwithstanding the negative and indifferent press from leading news outlets, Phoenix's clear emotional willingness to invest his visibility and solidarity capital strengthened his own brand and standing as an activist during a moment of professional climax. In other words, his speech was delivered to the world at the peak of a professional career in which he was being honored by audiences and peers, reflecting high favorability. A notoriously private person, the details that are publicly revealed about his personal life are consistent with his vegan lifestyle and altruism. Unlike the criticism surrounding his work (such as horseback riding), celebrity lifestyle choices fall under a form of public scrutiny that values authenticity without necessarily factoring in extenuating professional circumstances. A picture taken in the Netflix afterparty of him and fellow actor, vegan and partner Rooney Mara casually eating plant-based burgers went viral following the Oscars. Two days after the ceremony, with the image of the kidnapped calf and anguishing mother from his speech still fresh in the media, he joined Farm Sanctuary and Los Angeles Animal Save to rescue a real cow and her baby from slaughter, as reported by major outlets such as CNN, CBS, *The Guardian* and *The Independent.* The short documentary *Liberty* (2020a), once again directed by Shaun Monson, followed Phoenix in the rescue. That same year he again teamed up with Monson and with the environmental protest group Extinction Rebellion in the fictional short *Guardians of Life* (2020b), a film that allegorized a sickened planet Earth as a patient crashing in an emergency room.

The critical acclaim that has followed Phoenix for most of his career and that culminated with the Academy Award can be read as a sign of high public favorability of the actor. So too is his signature demeanor at festivals, ceremonies and interviews endearing to fans who post comments empathizing with his visible awkwardness and discomfort at being the center of attention. It does not go unnoticed how his initial embarrassment when walking upstage to receive his Oscar turned into confidence when he shifted the attention to nonhuman animals. Perhaps this was a calculated, performative act, yet audience expectations of moments of award reception as glimpses of live authenticity invest his behavior with credibility. Authenticity as an activist, strong favorability as an actor and likeability as a celebrity are ideal ingredients to sustain a movement's visibility and a prosperous Hollywood career. Parkinson and Herring note that Phoenix contradicts the claim that today's celebrities require a social media presence to validate their activism as authentic: "Phoenix's activism is, by comparison, private, to the extent that he does not publicise his plans or activities" (2023, 589). Even if this is a rare and unconventional form of celebrity activism, Phoenix's avoidance of personal social media profiles presents additional evidence to support his image as a highly private individual – an image that is consistent with his uneasiness at award shows and red carpets. This reinforces his authenticity as a personality and also trickles into perceptions of his activism as genuine. After all, "authenticity can be part of a celebrity's brand" (Farrell 2020, 7).

Interest in Phoenix's affective response to social issues has continued: in September 2025, there appeared images of him and Mara visibly moved after the premiere of *The Voice of Hind Rajab* (Ben Hania 2025) at the 82nd Venice Film Festival. Both actors served as executive producers of the film, which dealt with the violent killing of a Palestinian child in the hands of Israeli forces during the invasion of the Gaza Strip. At a time of intense international debate and a divisive political climate, the interest of news outlets and (social) media in showing Phoenix's unfiltered, heartfelt reaction can be read in several ways: (1) the images satisfied public consumption of celebrity intimacy; (2) the images promoted the film; (3) the images strengthened perceptions of Phoenix as authentic; and/or (4) the images found in Phoenix – fully equipped with the credibility that comes with positive perceptions of authenticity – a moral compass at a time when some nations were beginning to call the Gaza conflict a genocide. In this light, Phoenix's distress can be regarded as a symbol of grieving humanity. But then again, the images also threatened to distract from the political message of the film (and the conflict) in yet another instance of the "vampire effect."

In any event, the longevity of Phoenix's career suggests not just mastery of the craft but also savviness in the entertainment industry. His appearance in Casey Affleck's mockumentary *I'm Still Here* (2010), for the filming of which he publicly claimed to have retired from acting to become a hip-hop artist, evinced his knowledge of how the celebrity apparatus functions, how audience expectations are raised and how celebrity commodification can become quite profitable. Considering the film's staging of his meltdown, public forgiveness after what appeared to have been the parading of genuine erratic behavior for more than a year is of interest here. Nina Martin claims that, unlike his female counterparts, Phoenix had no problem reclaiming his public approval after the premiere of the mockumentary: "The masculine privilege, along with the fact that Phoenix was only pretending to unravel, allows him to theoretically put on a fat (sloppy and ugly) suit for eighteen months, and shed its negative effects immediately afterward" (2015, 30–31). I would add that in the case of Phoenix, public forgiveness for his behavior and his attempt to fool audiences was also due to his reputation as an actor who fiercely commits to his characters, *authentically* delivering outstanding performances. The post-premiere revelation that Phoenix's behavior had been scripted and planned was interpreted as a sign of artistic commitment at best and an eccentric celebrity whim at worst; nevertheless, his career remained unaffected and he went on to take on roles that continued earning him critical acclaim and ensuring professional success.

The celebrity embodied by Phoenix in the 2000s, therefore, is an intertext that both feeds from and nurtures his activist persona. His presence and name bring to animal advocacy documentaries a visibility that is not (and cannot be) detached from his reputation as an actor and a personality, which pours into his credibility as an activist. It is with an understanding of the intersecting affective responses to his acting, his persona and his activism that his framing of consensualism in these films ought to be approached. As an example, let us consider his contribution to *Earthlings*. Unlike the disclaimer in *Unity*, *Earthlings* does not mark a distance between narrator and script. Phoenix's voice, tone and cadence absorb Monson's writing, dismissing suspicions about a lack of correspondence between what the narrator says and how he truly feels. The original poster for the film shows half of Phoenix's face as the third part of a triad – Nature (a leafed plant) / Animals (a cow) / Humankind – that emphasizes connection and intersection. Phoenix's face is decentered to favor the visual presence of the cow in the middle. This is as far as he will show his face, precluding its threatening presence as a vampiric index.

The contents of the script he narrates foreshadow his imagery and wording in his Academy Award speech fifteen years later, addressing themes such as the correlation between racism, speciesism and sexism, the common capacity to suffer, and the moral obligation of humans to nonhuman animals. The first minutes of *Earthlings* are particularly insistent on the racism–sexism–speciesism analogy, a parallelism that, as discussed in Chapter 3, has a history of inviting criticism, especially when purported by privileged white males. Although public resistance to the analogy suggests that it may not be the best strategy to inspire concerted empathy, Phoenix's subsequent call for individual consumer responsibility does, on the other hand, align with the type of ideology championed by Western capitalist economies (best represented by the United States and its cultural empire). This appeal can be regarded as a form of consensualism insofar as it meets audiences at a middle ground where they recognize and feel comfortable with the cultural value of the individual as an agent of change, capable of impacting the economy and overthrowing tyranny. The moral stand of the (hyper-)individual, Phoenix suggests, is defined by consumer choices; by adopting forms of ethical consumption, one is converted and elevated into a higher form of being. Viewers hear statements in the line of: "[t]he time has come for each of us to reconsider our eating habits, our traditions, our styles and fashions, and above all, our way of thinking"; people manifest "a desire not to know about facts that may lie heavy on one's conscience"; "those who, by their purchases, require animals to be killed, do not deserve to be shielded from this or any other aspect of the production of the meat they buy." Even though the industries and practices that kill nonhuman animals by the millions are shown and described, the film promotes a form of change that begins with the individual, rather than delineating a top-bottom revolution that starts by targeting institutional and corporate conscience. This empowerment of the individual-as-consumer is, of course, part of the general framing of activist messages that claim that lifestyle is an important weaponry against corporativism. It is also a discourse that reproduces American mythologies of individualism and the promise of redemption (see Chapter 3). In this sense, Phoenix's advocacy in *Earthlings* is not really a defiance of the system that keeps the animal-industrial complex in place. As Brockington claims, "[c]elebrity activism for humanitarian causes tends to serve existing configurations of power, perpetuating existing relations, not challenging them" and revealing itself as "a profoundly pro-establishment political enterprise" (2021, 258). This is not to say that *Earthlings* lacks a critique of institutions, for practically every image of nonhuman animal suffering bears the weight of endless lists of victims of industrial complexes (see Chapter 1); however, against this machinery (and as a solution to it), Phoenix and Monson call on the individual. And if the individual is responsible for change, then he is also to blame for contributing and consenting to the system of exploitation in the first place. Invoking individualism, therefore, is a strategy of consensualism because it reproduces an argumentative logic that American/Western audiences are familiar with and for the most part, ideologically live by. This reduces the risk of feeling alienated by the celebrity; the viewer finds patterns of affective commonality with him, and experiencing this positive emotion is certainly helpful in the context of a film as brutal as *Earthlings.* Of course, celebrities can pull off discourses of meritocracy and individualism better than most people, as their personality brands are often already built around narratives of personal growth and perseverance that transform them into hyper-individuals that sell and regurgitate the rhetoric of the self-made man.

Consensualism is also discerned in Phoenix's narration and representation of exploitation in non-Western locations. At first glance, *Earthlings* appears problematic and confrontational in this regard; in particular, because of the long sections on dolphin hunting in Japan and the sourcing of leather in India. The violence with which dolphins and cows are treated contributes to a brutalized impression of both countries. Singled-out, these sections reproduce the paradigm of the Western "us" versus the othered "them" described in the previous section. However, hard as it may be to watch in one sitting, the film asks of viewers to engage with connections, including geographical ones. The narrator presents leather as a product that is primarily in demand in the US, Germany and the UK, thus placing responsibility (and blame) on Western consumers. He is also as critical of Western animal agriculture as he is of the Japanese fishing industry. There is no shying away from Western responsibility, which allows Phoenix and Monson to dodge accusations of racism. While, in a way, the myth of the great white savior can be regarded as a form of consensualism insofar as it reproduces a trope that millions of Westerners recognize and have been hardwired to identify with, the film's apparent rejection of that trope (even if Monson and Phoenix, both white males, stand as ideologues) nonetheless also entails consensualism because the script insists on a spiritually higher form of relating to the cultural other. This geographical, intersectional mode of conceptualizing animal exploitation is consistent with animal rights positions that seek reform in all avenues of domination based on hierarchical binarisms. Phoenix's "voice-of-God" authority advocates communal compassion and forgiveness. Years later, in his Oscars speech, he professes that forgiveness will elevate humankind. The Christian undertones implore an empathy toward the cultural and species other, again surrendering a rhetoric of deliverance that viewers recognize and have been raised to believe in.

Conclusion

Twenty-first-century media and pop culture marketing have created systems of celebrity engineering and consumption in which advocating for social movements maximizes a celebrity's ubiquitous presence and moral authority. If handled successfully (which depends on positive perceptions of the celebrity's authenticity, credibility and legitimacy), the symbiosis offers great benefits for both the movement and the celebrity. On the one hand, the alliance can grant animal advocacy a level of exposure that pulls it out of marginality, reaching a larger audience and allowing it to become more mainstream. The participation of celebrities as narrators, onscreen presenters and/or subjects in documentaries epitomizes a common attempt to "normalize" ethical veganism. On the other hand, when activist celebrities and celebrity activists invest their visibility and solidarity capital in animal advocacy, this can strengthen their public approval and protect their brand.

The partnership, nonetheless, is not without risk for either party. Celebrities can distract attention from the ethical message purported by films; the instability of their favorability can negatively impact the movement should any scandal or reason come to light to discredit them; their limited knowledge of the specific mechanisms and conditions of certain practices can be weaponized against them and expose them for their lack of institutional power; and perceived inconsistencies between their lifestyle and their message can erode their credibility. Moreover, their very nature as commodities is rendered possible by the same type of capitalist apparatus that the animal rights movement otherwise seeks to dismantle, and what was originally envisioned as a means to de-marginalize and shed negative vegan stereotypes can eventually be repackaged into profitable Big Vegan corporativism. When deploying an alliance between a celebrity and the movement, animal advocacy documentaries have to tread paths fraught with potential challenges posed by the celebrity as intertext.

Animal advocacy documentaries offer entertainment industry stars a space they can feel comfortable in: film. But to maximize the possibilities of message effectiveness, the celebrity must not only suffuse the film with positive perceptions of their authenticity and likability but also explore forms of consensualism that impede or mitigate the audience's sense of alienation from the subject matter. Having explored celebrity typologies and capital and how documentaries have typically instrumentalized their persona, this chapter has then approached the figure of Joaquin Phoenix as a case study showing how a celebrity who enjoys a generally favorable reputation engages with forms of consensualism in these types of films. Positive perceptions of Phoenix's authenticity as an activist enable viewers to consume the actor's rendition of the narrative script as a way to bond with him and the nonhuman victims on screen. It is hard to say whether *Earthlings* would have had the same impact had Phoenix not been involved. Nevertheless, the amount of affective capital that his celebrity generates and that is the object of media interest and scrutiny would suggest that the visibility he brings is considerable, which is reason enough to continue encouraging his involvement.

References

Affleck, Casey, dir. 2010. *I'm Still Here*. Magnolia Pictures.

Audiard, Jacques, dir. 2018. *The Sisters Brothers*. Annapurna Pictures, Why Not Productions.

Ben Hania, Kaouther, dir. 2025. *The Voice of Hind Rajab*. Willa.

Bieniek-Tobasco, Ashley, Sabrina McCormick, Rajiv N. Rimal, Cherise B. Harrington, Madelyn Shafer, and Hina Shaikh. 2019. "Communicating Climate Change through Documentary Film: Imagery, Emotion, and Efficacy." *Climatic Change* 154: 1–18. https://doi.org/10.1007/s10584-019-02408-7.

Blake, Katie, Sean C. Anderson, Adam Gleave, and Diogo Veríssimo. 2023. "Impact on Species' Online Attention when Named after Celebrities." *Conservation Biology* 38 (2): e14184. https://doi.org/10.1111/cobi.14184.

Boorstin, Daniel J. (1961) 1992. *The Image. A Guide to Pseudo-Events in America*. Vintage Books.

Bourdieu, Pierre. 1986. "The Forms of Capital." In *Handbook of Theory and Research for the Sociology of Education*, edited by John G. Richardson, 241–58. Greenwood.

Boykoff, Maxwell T., and Michael K. Goodman. 2009. "Conspicuous Redemption? Reflections on the Promises and Perils of the 'Celebritization' of Climate Change." *Geoforum* 40 (3): 395–406. https://doi.org/10.1016/j.geoforum.2008.04.006.

Brockington, Dan. 2008. "Powerful Environmentalisms: Conservation, Celebrity and Capitalism." *Media, Culture & Society* 30 (4): 551–68. https://doi.org/10.1177/0163443708030004 0701.

Brockington, Dan. 2009. *Celebrity and the Environment: Fame, Wealth and Power in Conservation*. Zed Books.

Brockington, Dan. 2021. "Celebrity Advocacy." In *Routledge Handbook of Humanitarian Communication*, edited by Lilie Chouliaraki and Anne Vestergaard, 253–65. Routledge.

Brookes, Gavin, and Małgorzata Chałupnik. 2022. "Militant, Annoying and Sexy: A Corpus-based Study of Representations of Vegans in the British Press." *Critical Discourse Studies* 20 (2): 218–36. https://doi.org/10.1080/17405904.2022.2055592.

Campanella, Bruno. 2020. "Celebrity Activism and the Making of Solidarity Capital." In *The Political Economy of Celebrity Activism*, edited by Nathan Farrell, 19–34. Routledge.

Campbell, Joseph. (1949) 1968. *The Hero with a Thousand Faces: Bollingen Series*. Princeton University Press.

Cherry, Elizabeth. 2024. "From Brigitte Bardot to Pamela Anderson. Celebrity Activists in France and the United States." In *Animal Activism On and Off Screen*, edited by Claire Parkinson and Lara Herring, 225–48. Sydney University Press.

Chouliaraki, Lilie. 2013. *The Ironic Spectator. Solidarity in the Age of Post-Humanitarianism*. Polity Press.

Cole, Matthew, and Karen Morgan. 2011. "Vegaphobia: Derogatory Discourses of Veganism and the Reproduction of Speciesism in UK National Newspapers." *The British Journal of Sociology* 62 (1): 134–53. https://doi.org/10.1111/j.1468-4446.2010.01348.x.

D'Ambrogio, Simone, Noah Werksman, Michael L. Platt, and Elizabeth N. Johnson. 2022. "How Celebrity Status and Gaze Direction in Ads Drive Visual Attention to Shape Consumer Decisions." *Psychology & Marketing* 40 (4): 723–34. https://doi.org/10.1002/mar.21772.

De Groeve, Ben, Liselot Hudders, and Brent Bleys. 2021. "Moral Rebels and Dietary Deviants: How Moral Minority Stereotypes Predict the Social Attractiveness of Veg*ns." *Appetite* 164: 105284. https://doi.org/10.1016/j.appet.2021.105284.

De Groeve, Ben, Daniel L. Rosenfeld, Brent Bleys, and Liselot Hudders. 2022. "Moralistic Stereotyping of Vegans: The Role of Dietary Motivation and Advocacy Status." *Appetite* 174: 106006. https://doi.org/10.1016/j.appet.2022.106006.

Delforce, Chris, dir. 2018. *Dominion*. Aussie Farms.

Doyle, Julie. 2016. "Celebrity Vegans and the Lifestyling of Ethical Consumption." *Environmental Communication* 10 (6): 777–90. https://doi.org/10.1080/17524032.2016.1205643.

Driessens, Olivier. 2012. "The Celebritization of Society Culture: Understanding the Structural Dynamics of Celebrity Culture." *International Journal of Cultural Studies* 16 (6): 641–57. https://doi.org/10.1177/1367877912459140.

Duthie, Elizabeth, Diogo Veríssimo, Aidan Keane, and Andrew T. Knight. 2017. "The Effectiveness of Celebrities in Conservation Marketing." *PLoS One* 12 (7): e0180027. https://doi.org/10.1371/journal.pone.0180027.

Erfgen, Carsten, Sebastian Zenker, and Henrik Sattler. 2015. "The Vampire Effect: When do Celebrity Endorsers Harm Brand Recall?" *International Journal of Research in Marketing* 32 (2): 155–63. https://doi.org/10.1016/j.ijresmar.2014.12.002.

Farrell, Nathan. 2020. "Introduction: 'Getting Busy with the Fizzy' – Johansson, SodaStream, and Oxfam: Exploring the Political Economies of Celebrity Activism." In *The Political Economy of Celebrity Activism*, edited by Nathan Farrell, 1–18. Routledge.

Fernández, Laura. 2021. "Images that Liberate: Moral Shock and Strategic Visual Communication in Animal Liberation Activism." *Journal of Communication Inquiry* 45 (2): 138–58. https://doi.org/10.1177/0196859920932881.

Frenette, Arielle, Caroline Desbiens, Julia Christensen, and Justine Gagnon. 2023. "Exploring Maternalism: Brigitte Bardot, Baby Seals, and Contemporary Hegemonies of the Female Hero." *Études Inuit Studies* 47 (1–2): 383–403. https://doi.org/10.7202/1113397ar.

Galkin, Matthew, dir. 2007. *I Am an Animal: The Story of Ingrid Newkirk and PETA*. HBO, Stick Figure Productions.

Giles, David C. 2013. "Animal Celebrities." *Celebrity Studies* 4 (2): 115–28. https://doi.org/10.1080/19392397.2013.791040.

Goodman, Michael K., and Jo Littler. 2013. "Celebrity Ecologies: Introduction." *Celebrity Studies* 4 (3): 269–75. https://doi.org/10.1080/19392397.2013.831623.

Heinich, Nathalie. 2012. *De la visibilité: Excellence et singularité en régime mediatique*. Éditions Gallimard.

Huggan, Graham. 2013. *Nature's Saviours. Celebrity Conservationists in the Television Age*. Routledge.

Ilicic, Jasmina, and Stacey M. Brennan. 2020. "Looking at You: Celebrity Direct Gaze Influences Social Media Post Effectiveness." *European Journal of Marketing* 54 (12): 3051–76. https://doi.org/10.1108/EJM-02-2019-0171.

Jain, Kokil, Isha Sharma, and Abhishek Behl. 2024. "Voice of the Stars – Exploring the Outcomes of Online Celebrity Activism." *Journal of Strategic Marketing* 32 (7): 842–63. https://doi.org/10.1080/0965254X.2021.2006275.

Jeffreys, Elaine. 2016. "Translocal Celebrity Activism: Shark-protection Campaigns in Mainland China." *Environmental Communication* 10 (6): 763–76. https://doi.org/10.1080/17524032.2016.1198822.

Jiménez-Morales, Manel, and Marta Lopera-Mármol. 2022. "Greening the Celebrification: The New Dimension of Celebrities through Green Production Advocacy." *Sustainability* 14 (24): 16843. https://doi.org/10.3390/su142416843.
"Joaquin Phoenix Wins Best Actor | 92nd Oscars (2020)." 2020. *YouTube* video, 5:41. Posted by *Oscars*, March 11. https://www.youtube.com/watch?v=qiiWdTz_MNc.
Jones, Huw D. 2022. "Achieving an Intimacy of Knowledge and Effect? The Impact of Documentary Films in Europe." *Journal of Audience and Reception Studies* 19 (1): 177–98.
Kahn, Rachel, and François Chaumont, dirs. 2019. *Brigitte Bardot, Le Serment fait aux Animaux*. Et la suite productions, France Télévisions.
Karlin, Beth, and John Johnson. 2011. "Measuring Impact: The Importance of Evaluation for Documentary Film Campaigns." *M/C Journal* 14 (6). https://doi.org/10.5204/mcj.444.
Katz, Brandon. 2020. "The 92nd Academy Awards Draw All-Time Low in Viewership." *Observer*, February 10. https://observer.com/2020/02/oscars-2020-tv-ratings-academy-awards-abc-viewership-parasite.
Kelly, Sean P., Gareth G. Morgan, and Tracey M. Coule. 2014. "Celebrity Altruism: The Good, the Bad and the Ugly in Relationships with Fundraising Charities." *International Journal of Nonprofit and Voluntary Sector Marketing* 19 (2): 57–75. https://doi.org/10.1002/nvsm.1487.
Kelner, Vincent, dir. 2022. *A Taste of Whale*. Warboys, Films Boutique.
Kossakovsky, Victor, dir. 2020. *Gunda*. Sant & Usant, Louverture Films.
Lockwood, Alex, dir. 2022. *The End of Medicine*. A. U. M. Films & Media.
Loy, Loredana. 2024. "About a Speech. The Media's Reaction to a Celebrity Advocacy Intervention." In *Animal Activism On and Off Screen*, edited by Claire Parkinson and Lara Herring, 299–323. Sydney University Press.
Mammola, Stefano, Nathan Viel, Dylan Amiar, et al. 2023. "Taxonomic Practice, Creativity and Fashion: What's in a Spider Name?" *Zoological Journal of the Linnean Society* 198 (2): 494–508. https://doi.org/10.1093/zoolinnean/zlac097.
Martin, Nina K. 2015. "'Does This Film Make Me Look Fat?' Celebrity, Gender, and *I'm Still Here*." In *Star Bodies and the Erotics of Suffering*, edited by Rebecca Bell-Metereau and Colleen Gleen, 29–54. Wayne State University Press.
Marwick, A., and Boyd, C. 2011. "To See and Be Seen: Celebrity Practice on Twitter." *Journal of Research into New Media Technologies* 17 (2): 139–58. https://doi.org/10.1177/1354856510394539.
Maxwell, Richard, and Toby Miller. 2012. *Greening the Media*. Oxford University Press.
Maxwell, Richard, and Toby Miller. 2013. "Should We Care About Celebrity Environmentalists?" *Greening the Media* (blog), *Psychology Today*, March 18. https://www.psychologytoday.com/blog/greening-the-media/201303/should-we-care-about-celebrity-environmentalists.
Mercy for Animals. 2011. *Farm to Fridge*, video, 11:50. https://www.youtube.com/watch?v=THIODWTqx5E.
Meyer, David S., and Joshua Gamson. 1995. "The Challenge of Cultural Elites: Celebrities and Social Movements." *Sociological Inquiry* 65 (2): 181–206. https://doi.org/10.1111/j.1475-682X.1995.tb00412.x.
Miller, Toby. 2024. "Celebrity Activism. PETA, Pamela, Porn." In *Animal Activism On and Off Screen*, edited by Claire Parkinson and Lara Herring, 271–97. Sydney University Press.
Moby, dir. 2023. *Punk Rock Vegan Movie*. Little Walnut.
Monson, Shaun, dir. 2005. *Earthlings*. Nation Earth.
Monson, Shaun, dir. 2015a. *Unity*. Nation Earth.
Monson, Shaun. 2015b. "Q&A: 'Unity' Director Discusses Humanity's Future. Interview by Marcelo Gleiser." *Marcelo Gleiser* (blog), August 12. https://marcelogleiser.com/blog-archive/qa-unity-director-discusses-humanitys-future.
Monson, Shaun, dir. 2020a. *Liberty*. Nation Earth.
Monson, Shaun, dir. 2020b. *Guardians of Life*. Nation Earth, Amazon Watch, Extinction Rebellion.
Murphy, Patrick D. 2021. "Speaking for the Youth, Speaking for the Planet: Greta Thunberg and the Representational Politics of Eco-Celebrity." *Popular Communication* 19 (3): 193–206. https://doi.org/10.1080/15405702.2021.1913493.

Napoli, Julie, and Robyn Ouschan. 2024. "Through the Eyes of Non-vegans: Using Photovoice to Explore the Stigmatisation of Vegans." *Qualitative Market Research* 27 (5): 750–99. https://doi.org/10.1108/QMR-03-2023-0037.
Nguyen, Sophia. 2017. "A Harvard Sociologist Studies How Veganism Went from Tasteless to Trendy." *Harvard Magazine*, June 8. https://www.harvardmagazine.com/2017/06/the-rise-of-vegan-culture.
Olausson, Ulrike, and Ylva Uggla. 2021. "Celebrities Celebrifying Nature: The Discursive Construction of the Human-Nature Relationship in the 'Nature is Speaking' Campaign." *Celebrity Studies* 12 (3): 353–70. https://doi.org/10.1080/19392397.2019.1626749.
Olmedo, Alegría, E. J. Milner-Gulland, Daniel W. S. Challender, et al. 2020. "A Scoping Review of Celebrity Endorsement in Environmental Campaigns and Evidence for Its Effectiveness." *Conservation Science and Practice* 2 (10): e261. https://doi.org/10.1111/csp2.261.
Olmedo, Alegría, Diogo Veríssimo, Daniel W. S. Challender, et al. 2021. "Designing Celebrity-endorsed Behavioral Interventions in Conservation." *Conservation Biology* 38 (3): e14273. https://doi.org/10.1111/cobi.14273.
Parkinson, Claire, and Lara Herring. 2023. "Vegan Celebrity Activism: An Analysis of the Critical Reception of Joaquin Phoenix's Awards Speech Activism." *Celebrity Studies* 14 (4): 584–601. https://doi.org/10.1080/19392397.2022.2154684.
Pelau, Corina, Puiu Nistoreanu, Laura Lazar, and Ruxandra Badescu. 2022. "Celebrity vs. Product: A Neuroscientific Approach to the Distractors in Food Advertising for Sustainable Marketing." *Sustainability* 14 (19): 12768. https://doi.org/10.3390/su141912768.
People for the Ethical Treatment of Animals. 2009. *Glass Walls,* video, 12:45. https://www.peta.org/videos/glass-walls-2.
People for the Ethical Treatment of Animals. 2014. *Face Your Food,* video, 3:57. https://www.peta.org/videos/peter-dinklage-face-food.
Phillips, Todd, dir. 2019. *Joker*. Warner Bros. Pictures.
Phoenix, Joaquin. 2018. Interview by Will Ferrell. *Interview Magazine*, March 28. https://www.interviewmagazine.com/film/joaquin-phoenix-april-2018-cover-story-will-ferrell.
Phua, Joe, S. Venus Jin, and Jihoon (Jay) Kim. 2020. "The Roles of Celebrity Endorsers' and Consumers' Vegan Identity in Marketing Communication about Veganism." *Journal of Marketing Communications* 28 (8): 813–35. https://doi.org/10.1080/13527266.2019.1590854.
Pöyry, Essi, Matilde Pelkonen, Emma Naumanen, and Salla-Maaria Laaksonen. 2019. "A Call for Authenticity: Audience Responses to Social Media Influencer Endorsements in Strategic Communication." *International Journal of Strategic Communication* 13 (4): 336–51. https://doi.org/10.1080/1553118X.2019.1609965.
Psihoyos, Louie, dir. 2009. *The Cove*. Ocean Preservation Society, Diamond Docs, Skyfish Films.
Psihoyos, Louie, dir. 2018. *The Game Changers*. ReFuel Productions, Oceanic Preservation Society, Diamond Docs.
Quesada Baena, Luis, Alice Binder, Ariadne Neureiter, Melanie Saumer, and Jörg Matthes. 2025. "'Do You Practice What You Preach?' The Effects of Celebrities' Pro-Environmental Messages on Social Media on Young Adults' Pro-Environmental Behavior." *Young Consumers: Insight and Ideas for Responsible Marketers* 26 (7): 22–37. https://doi.org/10.1108/YC-01-2024-1966.
Quinn, Christopher, dir. 2017. *Eating Animals*. Handsomecharlie Films, Heretic Films, Big Star Pictures.
Raphael, Jackie. 2020. "Authentic Activism. Challenges of an Environmental Celebrity." In *The Political Economy of Celebrity Activism*, edited by Nathan Farrell, 134–50. Routledge.
Reuber, Heike, and Beate Muschalla. 2022. "Dietary Identity and Embitterment among Vegans, Vegetarians and Omnivores." *Health Psychology and Behavioral Medicine* 10 (1): 1038–55. https://doi.org/10.1080/21642850.2022.2134870.
Rojek, Chris. 2001. *Celebrity*. Reaktion Books.
Rutledge-Prior, Serrin. 2023. "Vegans and 'Green-Collared Criminals': The Depoliticization of Animal Advocacy in Public Discourse." *Polity* 56 (1): 91–117. https://doi.org/10.1086/727840.
Schonfeld, Victor, and Myriam Alaux, dirs. 1981. *The Animals Film*. Slick Pics International.
Scott, Ridley, dir. 2023. *Napoleon*. Columbia Pictures, Apple Original Films, Scott Free Productions.

Sexton, Alexandra E., Tara Garnett, and Jamie Lorimer. 2022. "Vegan Food Geographies and the Rise of Big Veganism." *Progress in Human Geography* 46 (2): 605–28. https://doi.org/10.1177/03091325211051021.

Simonson, Peter. 2001. "Social Noise and Segmented Rhythms: News, Entertainment and Celebrity in the Crusade for Animal Rights." *The Communication Review* 4 (3): 399–420. https://doi.org/10.1080/10714420109359476.

Spang, Friederike. 2024. "Tailoring to the Audience? On the Potential Harms of Message Framing in Vegan Activism." *Journal of Agricultural and Environmental Ethics* 38 (6): 1–16. https://doi.org/10.1007/s10806-024-09941-1.

Suchan, Casey, and Denis Henry Hennelly, dirs. 2019. *The Animal People*. Finngate, Seafox, Birdstreet Productions, Virgil Films.

Sullivan, Sian. 2011. "Conservation is Sexy! What Makes This So, and What Does This Make? An Engagement with *Celebrity and the Environment*." *Conservation and Society* 9 (4): 334–45. https://doi.org/10.4103/0972-4923.92151.

Thürmer, J. Lukas, Juliane Stadler, and Sean M. McCrea. 2022. "Intergroup Sensitivity and Promoting Sustainable Consumption: Meat Eaters Reject Vegans' Call for a Plant-Based Diet." *Sustainability* 14 (3): 1741. https://doi.org/10.3390/su14031741.

Van den Bulck, Hilde, Nathalie Claessens, and Koen Panis. 2015. "World Relations and Developments Issues: Framing Celebrity Philanthropy Documentaries." In *Celebrity Philanthropy*, edited by Elaine Jeffreys and Paul Allatson, 105–24. Intellect.

Whiteman, David. 2004. "Out of the Theaters and into the Streets: A Coalition Model of the Political Impacto f Documentary Film and Video." *Political Communication* 21 (1): 51–69. https://doi.org/10.1080/10584600490273263.

Winge, Theresa M. 2008. "'Green Is the New Black': Celebrity Chic and the 'Green' Commodity Fetish." *Fashion Theory* 12 (4): 511–23. https://doi.org/10.2752/175174108X346968.

6 The Anthropocene and Post-Postmodernist Ethical Self-Reflexivity

This final chapter focuses on how animal advocacy documentaries have confronted the geologic epoch of the Anthropocene and the sensibilities of twenty-first-century post-postmodernist art. I analyze Simon Amstell's mockumentary *Carnage: Swallowing the Past* (2017), Louie Psihoyos's *Racing Extinction* (2015) and Emilio Fonseca's *Salvaxe, salvaxe* [*Wild, Wild*] (2024) – British, American and Spanish productions, respectively – as films that idiosyncratically reconcile post-postmodernist ethical self-reflexivity (which surpasses the aesthetic confines of postmodernist self-reflexivity) with narratives of hope that contest the defeatism and despair provoked by global environmental catastrophes. Squeezing out narratives of hope in spite and in light of anthropogenic disaster is a challenge for the twenty-first-century documentary image. The films analyzed here, I contend, manage to do so in inventive ways that experiment with the potential of self-reflexivity to persuade and motivate viewers to act. The first section discusses the aesthetic transition between postmodernism and post-postmodernism, at the center of which lies the desire for truth and an ethical engagement with others, and explores the importance of instilling hope in an ecologically fallen world marked by climate crisis and mass extinction. I posit three ways in which the films engage with ethical self-reflexivity: through genre, through the trope of (digital) technology as theme, and through the trope of (digital) technology as form. I then discuss *Carnage* as a film that uses the mockumentary genre to link postmodernist sensibilities with post-postmodernist ones, allowing Amstell, a comedian, to place parody at the center of an ethical agenda. Although *Carnage* is not concerned with the Anthropocene as such, it does deal with structures of the animal-industrial complex that have precipitated and contributed to its dynamics – structures that are part of the fabric of the Western culture that Amstell satirizes. In the second section I approach the trope of technology and Akira Mizuta Lippit's (2000) concept of animal spectrality to examine the possible implications these epistemologies have on our understanding of cinematic, historic and geologic time and on the notion of interspecies encounter. *Racing Extinction* reproduces much of the logic of Psihoyos's previous work with *The Cove* (2009); however, its upgrading and upscaling of the theme of (digital) technology reconfigures the significance of the archival image as index. The archival image signals the mourning of species loss, but it can

DOI: 10.4324/9781032661445-7

also be weaponized to inspire and instill hope for a better future, positing positive aesthetic response to beauty as the path to transformative change. In the final section, I analyze *Salvaxe, salvaxe*'s ethical self-reflexivity by focusing on the trope of (digital) technology as form. The film epitomizes Fonseca's experimental style with cinematography and sound engineering, envisioning how an ethical interspecies encounter materializes in the age of the Anthropocene, and thereby proposing an alternative understanding of connectivity between beings. This encounter must be produced by an asynchronous occupation of spaces and mediated by the digital reproduction of image and sound.

Hope, Resistance and Genre

In their study of the transition between postmodernism and twenty-first-century post-postmodernism, Jan Alber and Alice Bell (2019) address the academic claims that postmodernism has driven itself to extinction, that it is "a thing of the past" (Hutcheon 2002, 165). Versus the postmodernist celebration of self-reflexivity and conviction that truth is relative, contingent and subject to the whims of discourse, twenty-first-century aesthetic tendencies manifest a desire to recover the centrality of sincerity, the "real" and ethical commitment. Twenty-first-century art does not negate postmodern assertions on structure and discourse; it erects itself in spite of them. The postmodernist delight in parody and pastiche and its playful antics with the mingling of fact and fiction celebrate self-reflexivity as an end in itself, whereas twenty-first-century art seeks a meaningful connection with the world. "The effect is one of self-reflexive ethics" (Alber and Bell 2019, 130) that goes beyond the acceptance of the deceitfulness of language. Twenty-first-century art mnemonically signifies that "as humans, we need to believe in certain principles while simultaneously (and self-reflexively) acknowledge their artificiality" (2019, 130–31).

This defiance of postmodernism marks twenty-first-century art and is particularly pertinent in the context of the discursive apocalypticism surrounding the Sixth Mass Extinction, an effect of the current geological epoch. This era was proposedly christened the "Catastrophozoic era," "Homogenocene," "Myxocene" and "Anthrocene" (Kolbert [2014] 2024, 107), until finally landing on Paul Crutzen and Eugene Stoermer's (2000) nomenclature of the "Anthropocene," a term yet to be officially adopted by the International Commission on Stratigraphy (ICS). The term and concept of the Anthropocene, partly because of journalistic mishandling, partly because of a lack of unified discourses of scientific certainty, took a few years to sink into a mass media that was more attracted to only one symptom (global warming) (Sklair 2021) than to entertaining a more holistic apprehension of the problem from a geological perspective. The Anthropocene centralizes attention (and blame) on the noxious human impact on geology and ecosystems, with climate change, pollution and ocean acidification as top environmental concerns. The insidiousness of its effects has been described by Rob Nixon (2011) as "slow violence," that is, a type of violence that is gradual, even invisible, and dilated in time and space. Nixon states the need to understand that there is "a violence that is neither spectacular nor instantaneous, but rather incremental and accretive, its calamitous repercussions playing out across a range of temporal scales" (2011, 2).

Eco, animal advocacy and environmental justice documentary filmmaking in light of the Anthropocene seek ways to effectively represent those temporal scales and the interconnection of spaces. *An Inconvenient Truth* (Guggenheim 2006) and *Darwin's Nightmare* (Sauper 2004) are generally regarded as landmark films that, as part of the documentary renaissance of the early 2000s, traced the connections between globalization, capitalism, environmental degradation, poverty and political conflict.

Caught up in the industrial complex, nonhuman animals rendered as commodities had become the essential pieces of an insatiable apparatus dooming planetary survival. Food justice films in the line of *Food, Inc.* (Kenner 2008), *Farmageddon* (Canty 2011) and *Forks Over Knives* (Fulkerson 2011) chimed into the conversation by investigating the American corporate interests behind meat and dairy production – with *Forks Over Knives* clearly endorsing a plant-based diet. As discussed throughout this book, animal ethics issues acquired a sizeable presence in numerous films that examined exploitation and cruelty not as isolated matters, but as practices standardized by production systems, profitable commercialization and trafficking that cause environmental destruction and endanger underprivileged human populations. *The Cove* (Psihoyos 2009), *Cowspiracy: The Sustainability Secret* (Andersen and Kuhn 2014), *Eating Animals* (Quinn 2017), *The End of Meat* (Pierschel 2017), *Seaspiracy* (Tabrizi 2021), *Virunga* (von Einsiedel 2014) and *The Ivory Game* (Davidson and Ladkani 2016) are just a handful of examples. Undertaking a more global perspective, *Call of Life: Facing the Mass Extinction* (Thompson 2010) "urges its audience not to think of mass extinction just as a scientific or survival issue, but also as a spiritual and moral one" (Duvall 2017, 242). In a similar vein, *Eating Our Way to Extinction* (Brockway and Brockway 2021), narrated by Academy Award winner Kate Winslet, and *Racing Extinction* (Psihoyos 2015), discussed later, place extinction and loss of species biodiversity at the thematic center. These films' explorations of the systemic interconnectedness between animal exploitation and planetary degradation reframe and upgrade wildlife films' concerns with the disappearance of natural habitats and species – a theme that, according to film archivist Jan-Christopher Horak (2006), set the tone for the genre from the 1970s onwards. With the fear of extinction and the Anthropocene at the discursive center, urgency becomes heightened, and the world is presented as a ticking time bomb that requires immediate action, for humans can no longer lie back and enjoy the luxury of a generational postponement of responsibilities. Activist Greta Thunberg's often-quoted opening words to her 2019 speech at the European Parliament mirror the sobering tone of these anxieties: "I want you to panic. I want you to act as if your house was on fire."

Departing from postmodernism's general disengagement from sincerity and authenticity, twenty-first-century art produces narratives of resistance that inspire transformative change despite the sense of desperation and defeat caused by the apocalyptic scenario of the Anthropocene. The insistence on ethical commitment is often transfigured into texts of hope that urge political and institutional reform and simultaneously appeal to personal lifestyle changes to reduce one's carbon footprint. The fact that a digital publisher such as The Dodo has successfully expanded its brand into other media forms (for instance, a TV series on Animal Planet) reflects the viral, massive public interest in ethical and uplifting images of interspecies encounters. The Dodo's hallmark videos are rescue narratives: heartfelt (and melodramatic) stories of recovery with happy endings that trigger positive emotional responses and suggest that every person can be a hero to a nonhuman animal. This democratization of humane power is fueled by positive, optimistic and hopeful messaging, not by narratives of irremediable loss. Several scholars have described a moral obligation to create the circumstances that allow the seed of hope to grow. Often cited as an ethical response to the Anthropocene, Deborah Bird Rose's "connectivity ethics" subvert Western dualisms in favor of interspecies encounters premised on the subjecthood of the other. Connectivity ethics "are open, uncertain, attentive, participatory, contingent. One is called upon to act, to engage in the dramas of call-and-response, and to do so on the basis of that which presents itself in the course of life" (Rose 2011, 143).

Film Studies have also witnessed a growing interest in how to generate and sustain positive feelings by proposing constructive ways for humans to conceive and interact with the environment and other species. Alexa Weik von Mossner (2025) studies how nonfiction films about climate change articulate hope at the generational borderline between those who caused global warming and those who are expected to fix the problem. When films engage with discourses of hope, optimism and transformative agency, she argues, positive messaging can surpass the counterproductivity of despair. Chia-ju Chang (2017) also notes that film can offer a way to both redeem its complicity in the history of animal suffering and rendering and inspire hope for humanity in a time of planetary crisis. Chang focuses on documentary strategies that steer audiences' affective response away from the shock produced by images of violence (see Chapter 1 for more on moral shock). These strategies focus instead on inspiriting rescue activism, transforming narratives of cruelty into parables of redemption, or reframing animal subjects or species as important indexes in humans' existential exegesis of historic and geological time. For instance, she refers to the representation of the horseshoe crab as a "deep-time creature" whose ancientness in the oceans and in the stories of early human cultures is a claim to its status as "storied matter" (2017, 108) that prompts discussions about the current ecological crisis.

The films discussed in this chapter convey messages of hope by employing different modes and techniques that affectively entangle viewers and catalyze transformative action. As twenty-first-century post-postmodern art, they not only draw forth positive beliefs in the face of ecological collapse but also give rise to an ethics of aesthetic self-reflexivity that is proof of the decline of postmodernism. Postmodern art and criticism cannot hold their ground in a space and historic time marked by slow violence and human extermination of other species. Postmodern discourses and metalanguages become worn out by their own counterproductivity in the face of injustice, suffering and disappearance. Post-postmodernist sensibilities, however, can rehabilitate art's indebtedness to the human impulse to create a better world by reexamining the significance of self-reflexivity. When environmentalist author Bill McKibben published *The Age of Missing Information* in 1992, the antithetical separation between television/media culture and the "real," natural world seemed irreconcilable: screens brought an excess of information that led to the illusion of knowledge and the false promise of wisdom-within-reach through no other deliberate action than sitting in one's living room consuming television. Nature, on the other hand, connected the human subject to something real, granting access to a deeper understanding of life. It brought us face to face with our epistemological limits, which nurtured our wonder, and taught us to be at peace with them. But the multiplicative ontologies of mass media in the twenty-first century, propelled by the Internet, eventually redefined self-reflexivity in ways that demanded a reevaluation of nature and nonhuman animals inhabiting screens, which had become sites for the imaginings of posthuman geographies.

The marriage of digital imagery and the Anthropocene – and its derivatives, the "Plasticene" (Rangel-Buitrago et al. 2022) and the "Waste-o-cene" (Armiero and De Angelis 2017; Armiero 2021) – in nonfiction film precipitated a response that overcame the antinomy between the virtual and the real. By strategizing about the digital and the virtual as makers of subjects and meanings, post-postmodernism weaponizes the semiotics of metalanguage to devise structures and narratives that confront the Anthropocene and the discourses that deny its existence and/or severity. As such, the realist tradition in cinema that suggests that seeing amounts to knowing is upheld; but so is McKibben's ([1992] 2006) inference that seeing (media) equals misinformation. Twenty-first-century post-postmodernism points to self-reflexivity as a site for ethical knowledge, artificially and unapologetically manufacturing a mode of seeing that fosters hope as opposed to disillusionment, resistance as opposed to defeatism. Complementing Chang's (2017) and Weik von Mossner's (2025) studies on nonfiction film's proactive stances against the Anthropocene, what I propose in this chapter is the consideration of three other strategies that have been undertaken by filmmakers and that align with the moral compass of post-postmodernist aesthetics by engaging with ethical self-reflexivity: (1) the metadiscourse of genre; (2) the trope of (digital) technology as theme; and (3) the trope of digital technology as form. I analyze genre in the remainder of this section and discuss the other two strategies in the following ones.

In genre, postmodernism discovered a site for malleability, intertextuality and hybridity. For Linda Hutcheon (1988, 2002), genre could be employed for political critique, as it had the potential to question the authority of ideologies and institutionalized narratives and to fuse the historic with the fictional. Fredric Jameson (1991), on the other hand, viewed the commodification of genre as a reflection of the logic of late capitalism that pointed toward a form of self-reflexivity concerned with style over actual political criticism. Post-postmodernism overcomes genre's exhaustion with its own self-referentiality by infusing it with the possibility to channel authenticity, sincerity and ethical values. Referring to the cultural shift to metamodernism, Timotheus Vermeulen and Robin van den Akker state that

> new generations of artists increasingly abandon the aesthetic precepts of deconstruction, parataxis, and pastiche in favor of *aesth-ethical* notions of reconstruction, myth, and metaxis. These trends can no longer be explained in terms of the postmodern.
>
> (2010, 2)

Post-postmodernism (or metamodernism) thus utilizes genre as a mode to articulate hope and treats self-referentiality not as an end in itself but as a means to tap into the realness sought by new artists and authors beyond linguistic representation and textual layers (McLaughlin 2012).

In line with this shift in the significance of genre's self-reflexivity, animal advocacy documentaries have explored narrative structures that manage to address audiences' ethical concerns precisely through a semiotic system that signifies on concepts such as artificiality, representation, reproduction, distortion, parody and pastiche. Simon Amstell's mockumentary *Carnage: Swallowing the Past* (2017), a BBC production, stands out in this regard, and is in fact described by Alber and Bell (2019) as exemplary of twenty-first-century art's overcoming of postmodernist tendencies. Amstell, a comedian and writer, ventures into the mockumentary by adopting elements of speculative and futuristic fiction, blending them with documentary conventions that inform the genre's claims to truth. The formula is then inoculated with a heavy dose of parodic humor. The film's strategic inclusion of real footage of animal farming, images of the bovine spongiform encephalopathy (BSE – also known as "mad cow disease") breakout, and clips of current pop culture referents (such as celebrity chefs), pins the satire to real-life barbaric practices, creating a disquieting effect and preventing the film from being dismissed as mere satire. "This mockumentary is not playful for the sake of playfulness. Instead, it alerts us to the political, economic and moral issues that it humorously and often shockingly raises" (Alber and Bell 2019, 129).

Carnage is set in Britain in 2067, in a fictional context in which ethical veganism is now the dietary norm – the unmarked ethos. The film resorts to a narrator (Amstell himself) to purposely reproduce the type of male voice-over that is characteristic of the traditional expository mode. It documents the historical transition from carnism – presented as a grotesque, repulsive and cruel practice that now qualifies as the marked (and stigmatized) category – and the dissonance between younger generations and the residual bearers of national shame, that is, the older generations who once enjoyed meat and dairy products. Neither group is free from parody: the younger generation is portrayed as cringingly "soft" and ethereal, a hyperbolic display of the stereotype of vegans as socially accommodated and hand-holding, oversensitive do-gooders that are brought to tears at the mere thought of there once having been a time when carnism was the general diet. "Why would anyone eat a baby," one of them asks. "It's just a little baby. A little baby lamb." Matthew Adams regards *Carnage*'s humorous portrayal of vegans as an act of "self-deprecation" that "bypasses audience tendencies to trigger exaggerated and polarized representation of the 'other' as defence mechanisms, conscious or otherwise, when difficult or taboo issues are raised" (2022, 133). By employing parody, Amstell appropriates and secures his control of the representation of vegan youths, and ensures that viewers perceive the film as fair, insofar as it equally satirizes all social groups. The past, the narrator informs, has been masked by a tacit silence, a sign of the shame held by older generations over their former complicity as consumers. In a smart twist of wildlife documentary's archivistic framing of its content as possibly the last images of near-extinct species (Horak 2006), older generations are ethnographically pursued as indexes of an unforgivable past, the last historical witnesses of a time marked by carnism and the surrounding triviality of mainstream media and pop culture. Drenched in guilt, some of these humans form weekly support groups where they verbalize their past actions – including the naming of a type of cheese they once ate – to confront their trauma and heal.

Mockumentary frippery allows Amstell to parody all aspects of human historical behavior. It is not so much the Anthropocene as a geological epoch that is subjected to parody; mockery is instead directed against the media's levity regarding global warming and climate disasters, and against cultural degeneracy and consumerist debauchery. *Carnage* unleashes its humorous critique on the meat industry's advertising and campaigning, celebrity culture, the fashion industry and climate change negationists, but also makes room to hyperbolize the discourses of academics, animal activist stunts and performances. Technological devices that are the commodities of 2067 look absurd but aptly match current consumer desires (for instance, virtual realities and floating chairs). The fact that a machine built in 2032 is finally able to translate nonhuman animal thoughts and emotions into human language reinstates thc long-standing conviction of animal advocates that speciesism in the Aristotelian tradition (based on humans' distinct possession of the faculty of logos and reasoned response) is what has largely allowed animal exploitation to persist. This speaks to the film's concern with language at a grander, thematic scale: the younger generation lives by an ethical and lawful practice so widespread that the term "veganism" has fallen into disuse; the past has become too traumatic to relive with words; carnism is bluntly redefined as a diet of dead matter and stolen fluids; and gendered pronouns are employed to refer to animal subjects.

Carnage's parodic portrayals deform the logics of Carl Plantinga's "spillover effect" (2009, 184–87), which establishes that viewers' feelings of pleasure result from the physiological carryover of the negative emotions they have previously experienced. In the film, the arousal of negative emotions through distressing archival images and the reframing of pop culture referents as grotesque sensationalists is refracted by the absurd, ritualistic enactment of the older generation's support group. The parodic mode interrupts the relief and closure which positive emotions, as manifestations of catharsis, tend to produce. Instead, Amstell's formula – with a temporal setting in which the past is our present and the present is our future – suppresses the sense of peace and resolution that the younger generations, as representatives of positive affect and happiness, are meant to convey. Indeed, one of the things that the mockumentary genre prevents in *Carnage* is the sense of closure: the younger generation, as a model of social outcome for humans (however implausible by today's realistic standards), does not, because of their exaggerated, soft-lensed and warm-lighted serenity and intimacy, inspire serious imitation. Amstell's sarcastic deformation not of vegans themselves but of the stereotypes embodied by these fictional characters impedes a "proper" resolution of the spillover effect that may reconfigure the residue of negative emotions. And by interrupting the formula of antagonistic positive/negative units and preventing a sense of closure that invites the satisfied viewer to move on, he creates the circumstances that give us pause. Statements, images and enactments, cloaked by humor and the historical lens, reveal the absurdity and cruelty of current, real-life carnism, creating a new language with which to describe it.

When the character of Dr. Yasmine Vondeburg (played by English actress Linda Bassett) is introduced as an author exploring Britain's carnist past and helping other ex-carnists cope with their shame, she appears as an interviewee stating that "It's still difficult to say it out loud. 'I ate animals. I ate fish. I drank the milk from cows.'" She pauses for a moment, grimaces and repeats in disbelief, "From cows!" Her slight nod as she repeats this turns into subtle head shaking, configuring a visible manifestation of the human subject's shock and incredulity at their past actions. While the appropriation of conventions used in trauma-themed documentaries (a reliance on the interview form, a duration of shot sufficient to "read" the silences of interviewees, the discomfort of dealing with the ineffable etc.) triggers humor (the comedic effect being produced by the "laughable" speculation that eating nonhuman animals could someday amount to a Holocaust-like narrative authored by witnesses and former partakers), one cannot help but linger on Dr. Vondeburg's astonishment as she repeats "From cows!" The disbelief contained in that repetition is Amstell's way of inviting the viewer to actually and *really think* about it. The striking reminder that cows are robbed, as mothers, of the nourishment meant for their babies, converges with the visceral disgust of consuming another being's fluids. The positive emotion of compassion becomes intertwined with the negative feeling of disgust. Amstell circles back to this affective arousal in other moments, such as in the sequence about a 2024 hit musical broadcast by the BBC in which a female actress plays a dairy cow in a ridiculous costume and sings the following lyrics:

> Where? Where have you taken my baby? / Moo! He needs his mother too . . . I'm not a machine, I'm a mother / And my nipples they bleed / It's my blood in your milk / And there's pus in your milk . . . What kind of animal rapes just for milk?

Disgust here meets comedy, compassion and horror.

The film's fictional historical timeline is sequenced with the use of actual news and stories. In an impressive historiographic exercise, *Carnage* rescues actual reports, anecdotes and events to reconstruct a narrative of key moments in Britain's dietary reform. For example, the 2015 Piggate claim that Prime Minister David Cameron had performed acts of bestiality is briefly mentioned not in light of how it was actually depicted in the media (as a rumor seized as an opportunity for political advancement), but in light of one thought-provoking remark: "Somehow, placing one's penis in a pig was considered much worse than killing one and putting him or her in a sandwich." The film manages to inform the public of actual facts relating to the consumption and exploitation of nonhuman animals – facts that are spoken by a prophesized generation that has become historically equipped with a language that surpasses today's institutionalized speciesist bias. In one moment, fictional academic Maude Polikoff (played by Tony Award winner Lindsay Duncan) manages to simultaneously inform, provoke disgust and invite reflection by describing egg consumption in a manner so outrageous that it initially elicits laughter:

> I think people *did* know that an egg is a chicken's period. But why not have a period for breakfast? I don't think people knew that wild hens only have ten to fifteen periods a year. Imagine being forced to have twenty times more periods than you were supposed to and then being killed for lack of periods. It was beyond sexist.

The film's transgression of documentary and fictional structures allows for a reflexive contemplation of the mockumentary's form and content (Sánchez-Vilela 2019, 59) while presenting data about cruelty and exploitation by means of a language that foregrounds similes, parallelisms and semantic fields that reinstate the nonhuman animal subject. Hybridity is accomplished by blending conventions of the fictional and documentary signifiers – a pastiche that ultimately, following post-postmodernism, uses the semiotics of genre to call for responsible action and inspire hope. As Weik von Mossner notes, *Carnage*'s "actual purpose is to help contemporary viewers see their own normalized behavior from a different, unfamiliar angle" (2021, 315).

Carnage's explicit pastiche of myriad media genres (real and fictional interviews, news coverage and reports, film sequences, celebrity chef programming, advertising, talk and variety shows, radio shows, music, performance art, etc.), creates a collage circumscribed by the larger mockumentary genre. This is Amstell's formal proposal to make animal advocacy palatable for today's viewers. Amstell might well be hinting at his own self-reflexive endeavor in a sequence explaining fictional activist Troye King Jones's (played by actor John Macmillan) changes in his filmmaking. Having originally created videos aiming at moral shock (here Amstell takes the opportunity to show the brutal, massive killing of male chicks in the egg industry), Jones has transitioned to "lighter versions" that involve scripted videos with hand puppets, toy chicks and fake blood. As suggested earlier, *Carnage* manages to infiltrate real raw footage exposing the animal-industrial complex, but the mockumentary genre altogether constitutes a "lighter version" of what could have been constructed as another filmic anthology cataloguing animal suffering and aiming at moral shock through elements of horror, gore and tragedy (see Chapter 1). And it is probably because of the levity of the mockumentary genre that it was produced by the BBC in the first place. Because "[W]ho wants to sit and watch an entire film about veganism?" Jones asks, cluing viewers into Amstell's own synchronous accomplishment of having them watching and consuming *Carnage*. Jones's (and Amstell's) momentous comment stands at the cusp of mockumentary structure and serves post-postmodernist ethical self-reflexivity: the exposure of the genre's signifying games creates a mitosis between the implied viewers of the 2067 documentary and the real-world viewers of the 2017 mockumentary, and in that process offers a space for ethical contemplation of the viewer's responsibility as an agent of transformative change.

The Trope of (Digital) Technologies as Theme

The second strategy employed by filmmakers to create narratives of hope and post-postmodernist ethical self-reflexivity is the thematic reliance on the trope of (digital) technology. The onscreen representation of (digital) technology as an instrument that can transform optimism into hopeful action is a recalibration of postmodernism's subjections of that same technology to an exegesis that only flatters its capacity to imitate, recycle and delight in its own capacity to reproduce. Postmodern emphasis on the palimpsest of discourses that signify on reality has led to an assimilation of (digital) technologies as producers of likeness with the capacity to frame, represent, deconstruct, imitate and distort, stylizing multiple rhetorical devices (hyperbole, synecdoche, metonymy) to achieve this. Mass media and the omnipresence of the digital image saturate domestic and public spaces. Ubiquitous screen culture, soundbites and the packaging of thoughts, messages, opinions, sentiments and responses into consumable videos, memes, comments and tweets have facilitated the spread of fake news and alternative facts, creating a cultural scenario that distrusts traditional scientific, medical and intellectual authorities (Farkas and Schou 2020, 2) and instead favors the "authenticity" conveyed by unfiltered (and uninformed) statements, often enough influenced by the populism that underpins current Western politics. Technology, in other words, is what has made the crisis of credibility of the post-truth era possible. And while this crisis demands mining of adequate criteria to assess trustworthy sources (Petersen 2023), an interesting strategy undertaken by some filmmakers is to overtly address the distortive and imitative potential of digital technology, render it evident for the viewer and refuse to deny its capacity for sensationalism, with the ultimate aim of using those features to seek hope and truth.

In her book *Ecstatic Worlds: Media, Utopias, Ecologies* (2017), Janine Marchessault traces the epistemologies and phenomenologies of worldmaking in twentieth- and twenty-first-century photographic media and digital image. Exploring concoctions of ecotopian spaces and futures created by artists and filmmakers, Marchessault argues that new forms of knowledge enabled by technological supremacy over the human sensorium have the potential to "create new kinds of collectivity that are inclusive of a multiplicity of singularities, including the nonhuman." This invites the possibility to envision "more-than-human environments that break away from their anthropocentric origins to become 'things' not reducible to humans" (2017, 250–51). Marchessault's view suggests that aesthetic proposals interrogating anthropocentrism do not limit theorizations on (digital) technology to its capacity to copy, but instead conceive the imitative gesture as a means to create something beyond a reproduction, something that makes a statement about the moral state of affairs regarding species others.

I argue that the trope of (digital) technology, as theme and as form (the latter is discussed in the next section), has been used by animal advocacy filmmakers to confront the Anthropocene by proposing forms of species connectivity that supplant anthropocentric narratives. Notwithstanding stylistic dissimilarities, grammars of the mediatized image in these films exacerbate the world-making possibilities of digital reproduction. Rejecting the verisimilitude of realism and the limits of observational cinema, they create images of nonhuman animal cyborgs that, paradoxically, signify the factual nonhuman animal body. In the post-truth, post-postmodernist context, attention (and ethical response) to the nonhuman animal is accomplished by first framing the metadiscourse of the image. Focus is directed toward the indexes that remind the viewer that the image is, indeed, an image, a representation. Yet these indexes are then used to address the facticity of the nonhuman animal body, one that, in the age of the Anthropocene, spectrally moves between presence and absence.

At the beginning of the twenty-first century, Lippit (2000) theorized about the epistemological significance of the "disappearance" of nonhuman animals in the hands of capitalism. Drawing on John Berger's famous 1980 essay *Why Look at Animals?*, Lippit explores the relegation and confinement of nonhuman animals to postmodern technological spaces. Berger had characterized modernity as an era marked by the disappearance of nonhuman animals, in which capitalism and urbanization have transferred them to spaces where they are reduced to illusions rooted in cultural expectations that satisfy economies of consumption. Animals have become marginalized and "co-opted into the *family* [as pets] and into the *spectacle*" (Berger [1980] 2009, 25; emphasis in original). Their dislocation precludes any meaningful encounter, and as adulterated copies of the creatures they once were, they occupy the rims of existence as attenuated souvenirs of an original interspecies habitation of the world. The modern retrieval to the family unit and the private sphere conceives of pets as "mementoes from the outside world" (Berger [1980] 2009, 24), while nonhuman animals kept in zoos and forced into inaction and apathy become "the living monument to their own disappearance" ([1980] 2009, 36).

Lippit returns to Berger's historiography of loss and marginalization to study the displacement caused by the hyper-visibility of technology. This hyper-visibility and ubiquitousness, he argues, has paradoxically led to the invisibility of nonhuman animals, as they become signs and vectors designed to satisfy humans' explorations of themselves. As such – and in consistency with postmodern centrality in the constructedness of discourse – nonhuman animals are stuck in a perpetual state of vanishment: technology, as their new habitat, does not allow them to disappear altogether, restraining them in a constant state of becoming-absence, where they enter "a new economy of being" that is "*spectral*" and where they "[linger] in the world *undead*" (Lippit 2000, 1; emphasis in original). Technologies of reproduction and representation are historically bound to nonhuman animal bodies. Eadweard Muybridge's canonical studies reflect the interest that photography and the moving image have historically had in nonhuman animal motion. But, as Nicole Shukin (2009, 104–5) ascertains, it is also important to remember that nonhuman animal bodies were rendered to create the original gelatin used in photographic and film stocks. As trope and as rendered material, therefore, nonhuman animals have been forced to inhabit technoscapes that channel, disperse, repeat and multiply their bodies and movements and that distract from their extinction in the actual, real world. As Lippit states, "technology and ultimately the cinema came to determine a vast mausoleum for animal being" (2000, 187). The "technological crypt" (2000, 189) allows for images of nonhuman animals to be rewatched, replayed and reproduced, precluding the completion of their vanishment and falsely reinforcing an illusion of their ubiquitous presence in the world. I contend that these false impressions (or, one could argue, alternative facts) distort factual information about the dire situation faced by nonhuman species in the Anthropocene. At the same time, by keeping animals "undead," technology carves a space for their memory, becoming a "massive mourning apparatus" (Lippit 2000, 188). Cinema stepped into a fundamental role here, as it "provided an alternative to the natural environment that had been destroyed and a supplement to the discursive space that had never opened an ontology of the animal" (Lippit 2000, 196–97).

The question for animal advocacy filmmakers striving to create a message of hope and resistance is how to effectively employ technology as a recourse that can retrieve nonhuman animals from this cryogenic state and free them from a paralysis fixated on perpetual historic loss. The image must be one that empowers viewers with a belief in their own capacity to change the world; not one that dwells in an aesthetic contemplation of historic irreparability. Here begins the affirmation of a nonhuman animal ontology, one that starts with the ethical acknowledgment of the other as subject and co-sharer of a space and time defined by the Anthropocene, where all creatures' vulnerabilities come together. As thematic and formal self-reflexive tropes, (digital) technologies of reproduction are sought as ways of thinking about the Anthropocene and are weaponized as instruments of resistance.

Louie Psihoyos's *Racing Extinction* (2015) stands as a perfect example of (digital) technologies of reproduction employed as a thematic trope that is narrativized to inspire hope. The film sets out to educate about the scientific evidence and environmental circumstances causing the Sixth Mass Extinction and the Anthropocene. There is a clear activist agenda tied to the revelation of the *truth* behind human impact on the planet, ecosystems and biodiversity, with a straightforward plea for institutional change and the reevaluation and reform of personal lifestyle choices. The opening sequence foregrounds the film's preoccupation with the technological: a close shot of a jar containing the body of the last dusky sparrow, who died in 1987, sets the tone of death and disappearance. Behind the jar, a photographer (possibly Psihoyos himself) takes a picture of the jar, and the recording and photographic cameras mirror each other's archiving of the stillness of nonhuman animal death. As Psihoyos's voice-over presents the theme of extinction, shots of collections of dead nonhuman animals, body parts and remains, pinned or stocked together in shelves and storage rooms (some doubtless confiscated trophies), create an overwhelming feeling of loss and sadness, accentuated by the sorrowful piano tracks of Ralph J. and Anohni's (credited as Antony) "Manta Ray," the film's original song that would later be nominated for an Academy Award.

What begins as a graphic archive of the abundance of absence and loss becomes, throughout the film, contested by activists' attempts to restore species, rehabilitate habitats and in the process redeem humanity. Samples and conversations about technological devices (phones, specialized cameras, recorders, eco-friendly versions of cars, etc.) cohesively thread the narrative together. Time as captured by technology in a history of extinction becomes a determining factor in the conceptualization of the image of the nonhuman animal as archive of the dead or as compelling materiality of the still living, bringing the spectral to an impasse between always disappearing or being resuscitated into restoration. Belinda Smaill writes that

> onscreen encounters between life forms are produced in ways that rely on different modalities of time – the time of the narrative unfolding in shots, audio and edits; reception and the instant of bodily response to bodies onscreen; and the interval between capturing the image of life with the camera and projecting it on screen.
>
> (2016, 16)

Absorbed in a rhetoric of temporality, *Racing Extinction* thematizes the immediacy of digital technology to transform feelings of despair into action, for which the presence of the body is central. Versus the stillness of the dead nonhuman animal, technology offers movement and with it, the possibility of life. Movement signifies vitality, the active spatial occupation of the nonhuman animal body in time, surviving, carrying on. The cameras also capture the incessant movement of the activists, situating cinematic fixation with the embodiment of labor (Smaill 2016, 26) in the contextual urgency of the Anthropocene. PhotoArk, a project founded by National Geographic photographer Joel Sartore with the aim of photographing every live captive species on Earth to raise awareness about the critically endangered ones, is represented as more than an archive mourning loss. In the stillness of Sartore's work there remains life, and it is portrayed with beauty – an aesthetic means (but not an end in itself) to which the entire film aspires. As Psihoyos eventually states, "To me, photography isn't just about showing somebody how beautiful something is. It's a weapon." And as Sartore adds, "I really hope the PhotoArk isn't just some sort of an archive of the things we lost, but instead it's a chance to get people interested, look these animals in the eye and fall in love with them."

Racing Extinction in many ways replays the signature style of *The Cove* (see Chapter 3 for more on this film), in which Psihoyos establishes a dichotomy of human forces in the Anthropocene by remarking that "to me, you're either an activist or an inactivist." The two films follow similar patterns. Among other features, they both resort to: (1) the on-screen emphasis on activists as heroes; (2) their covert investigations in Asia reminiscent of spy films; (3) against-the-clock pacing;

(4) the authoritative support of interview material; (5) the exposure of business and political interests in the regulation and decision-making affecting animal welfare and conservation; and (6) the prominence of charismatic activists who embody hope despite the odds (the late primatologist Jane Goodall, whose writing and public appearances were markedly conceived around the need for hope, appears in *Racing Extinction*). Like *The Cove* as well, the promise of deliverance is encoded in the spectral presence/absence of nonhuman animals captured and compiled by (digital) technology. In *The Cove*, the activists' retrieval of images and soundscapes of the dolphin slaughter in Taiji narratively befits the culmination of the trials and tribulations endured by the heroes. The devices employed for such recovery are celebrated as technologies that create records and bear witness.

Significantly, and as discussed in Chapter 3, *The Cove*'s final gesture to inspire people is an appreciative nod to technological reproduction: a screen showing the violent slaughter is strapped onto the film's leading, charismatic activist, Ric O'Barry, who brings the images to the International Whaling Commission (IWC) and the streets of Tokyo. In *Racing Extinction*, the final climax imitates this cycle but with signal differences. Psihoyos goes "bigger," literally: by recruiting the talent of digital artist and visionary Travis Threlkel (founder of Obscura Digital), known for his large-scale projections on iconic buildings across the world, the "grand gesture" is devised. A Tesla Model S is customized with a 15,000-lumen projector, an electroluminescent paint job (modeled on bioluminescent organisms) and a FLIR (forward looking infrared) camera to visualize CO_2 emissions. The vehicle is then driven around to project images of wildlife on urban surfaces. The images are complemented by short informative textual messages about species extinction. Projectors are also set to screen large-scale images and messages on the surface of the United Nations Headquarters and the Empire State Building, among other sites. The projections on these Manhattan buildings were carried out almost a year apart (on September 20, 2014, and August 1, 2015, respectively), though the film editing of the final scene would appear to support the impression that they took place on the same climactic night to summon the cathartic powers of a narrative unity of time. Some of the projections show nonhuman animals and ecosystems in motion. Others create the illusion of movement with still images that slowly glide over the surface of the building, using material from Sartore's Photo Ark along with works by photographers Shawn Heinrichs and David Doubilet (Sartore 2015). Faces of people from across the world that are vulnerable to environmental injustice and political conflict are also displayed. Other projections represent the spectralization of wildlife as the images disintegrate into particles and, bloodlike or dust-like, trickle down the surface, enshrining technology with readable digital metaphors of species disappearance. In this manner, the liveness and movement accomplished by the digital copy – the cyborg nonhuman animal – contests the death and haunting stillness of the film's opening visuals of the bodies and dead matter that compound the archive of jars and storage rooms of irreparable loss. Digital motion resonates with the title's gerund, *racing*, which reinvigorates the idea of an ongoing commitment and struggle.

Tellingly, the projections are not meant to stun viewers into moral shock with graphic images of violence. Unlike the images of dolphin slaughter that O'Barry carries on his screen in *The Cove*, the digital projections of *Racing Extinction* display the wondrous beauty of nonhuman animal faces and bodies, especially when in motion (or simulated to appear as such). The compelling effect of the finale is enhanced by the film's sequential playing of J. Ralph and Sia's uplifting song, "One Candle," followed by "Manta Ray" in the final credits (this time with Anohni's rendition of the lyrics, which are written from the point of view of Mother Nature, or Earth, whose children are dying). While "One Candle" is written in a major key to exalt feelings of empowerment, "Manta Ray" is in the key of A-flat minor, creating a more somber soundscape. I would argue that the musical hopefulness of the first song has a spillover effect on the sorrowfulness of the second one, sustaining an emotional intensity gravitating around the understanding of the beauty of life. The film's strategy to visually and sonically overwhelm viewers with beauty echoes Martha Nussbaum's (2022) exaltation of wonder as a constructive emotion that can lead to transformative change (see Chapter 2). The beauty of an image is central to conservation efforts because of its potential to move viewers beyond mere contemplation, leading them into an affective state that will inspire them to act to preserve that beauty. Beauty is not exclusive to charismatic megafauna; as a recent study shows, edited images of species that are aestheticized to please the human eye provoke a "beauty bias" that positively impacts public attitudes and perceptions of that species (Shaw et al. 2024). The integration of beauty in discourses of hope in the Anthropocene appears in other recent documentaries that stand at a thematic intersection between environmentalism and animal advocacy, such as *My Octopus Teacher* (Ehrlich and Reed 2020) and *All That Breathes* (Sen 2022). Beauty of shot in all three films is considerably dependent on the affective value of the negative space and the compositional relationship it creates with the (non)human subjects – the kelp forest in *My Octopus Teacher*, the skies and rooftops of Delhi in *All That Breathes*, and the upwards-spiking jungle of New York skyscrapers in *Racing Extinction* are aesthetically pleasing landscapes that frame the species image. "My feeling is that by bringing nature to the city we're hopefully able to inspire people to actually help save these creatures," Psihoyos says. *Racing Extinction* shows a similar attempt in Lamakera, Indonesia, the ground zero of manta ray fishing, where activists set up a large screen for the locals to observe images of the beauty of their prey in consonance with the beauty of their village. "We want them to get an entirely new view that the stuff they go out in the water and they see every day and they stick a spear in is worth far more alive than dead," a conservationist states.

By naturalizing an ontological correspondence between nonhuman animals and their digital representations, *Racing Extinction* thus breaks with the conception of digital technology as a "crypt" of the (un)dead, instead absorbing and constructively building on the postmodern proximity to nature through the screen. But rather than dwelling on the technological copy as a language merely signifying on its own imitative capacities, it treats the beauty of the copy as an uplifting force from which to salvage hope, pushing the postmodern aesthetic toward a post-postmodern rationale. The verticality and symbolism of the New York buildings, a testament to the great wonders that humans can erect, is linked to the aesthetic pleasures aroused by the contemplation of wildlife, allegorizing a logic that ties humanity's capacity for greatness with nature's gift of flourishing biodiversity. All species are united against the common threat of the Anthropocene. The spectrality of the projections transcends the image as an object of mourning: beyond the grief of loss, there is the will to resist and to fight back.

The Trope of (Digital) Technologies as Form

The third strategy discussed in this chapter that complies with post-postmodernist ethical self-reflexivity is the trope of (digital) technologies as form. In this instance, technological self-referentiality is central to preproduction, principal photography and post-production, and the multiplicity of cinematographic possibilities is foregrounded in a filmic exercise of deliberate formal hyperbolism. Form need not seek beauty or immediate shock; as discussed later, it can be used to produce a disquieting effect, an uneasiness that separates the viewer from the comfort zone of aesthetic pleasure and stirs a sense of restlessness. This affective disturbance awakens viewers from their apathy toward the Anthropocene; as the image of the nonhuman animal is deconstructed and exposed as a discourse, the real threat of species extinction effectively centralizes moral response.

Emilio Fonseca's *Salvaxe, salvaxe* [*Wild, Wild*] (2024) is a unique documentary in this regard. The film revolves around the disappearance of wolves in the northwestern habitats of the Iberian Peninsula (the Spanish autonomous community of Galicia and northern Portugal). Like *Racing Extinction*, technology is also a central thematic trope, as the filmmaker and crew team up with conservationists and researchers to track the few scattered wolves in the territory, who are on the verge of extinction. The scientific technology that provides quantitative and qualitative data on the hidden presence of packs and lone wolves in the Galician forests is juxtaposed with the technical equipment that Fonseca employs for filming. The narrator intervenes in the form of carefully timed textual captions – Fonseca avoids the authoritative presence of a voice-over – and informs the viewer about this technology. But *Salvaxe, salvaxe* departs from *Racing Extinction*'s technological self-reflexivity by emphatically exploring digital reproductions of texture, light, sound and movement as storytellers themselves. In other words, Fonseca foregrounds form and cinematography to visibly remind the viewer of the artifice of the digital image. This auteurist endeavor is devised as a conscious creative step away from the conventional formal structures that appear in traditional conservation films, which are closer to the style of *Racing Extinction.* Not only does the dislocation of form situate *Salvaxe, salvaxe* in a more undefined category regarding genre (perhaps best described by terms such as "experimental documentary" or "film essay," as settled by IMDb; or "nature antidocumentary" and "cine-tracking," as stated in the film's press kit), but it also problematizes the spectral presence of the nonhuman animal in new ways.

Salvaxe, salvaxe resorts to the semiotics of an array of visual and sonic mechanisms that point to their very artificiality and displaced relation to time. Camera-trap footage of wolves and other wildlife is manipulated using close-ups, frame enhancement, looping of movement and frame-rate alterations; zenithal shots of ground textures, pawprints and trails (using satellite images) are conspicuously enduring; handheld camerawork registers the abruptness and unevenness of strenuous trails, with the camera becoming another body subjected to movement, light and temperature; haptic shots detail with sensitivity the density of soils, vegetation and the skin and hair of dogs and humans (the scientists); there are eye-level shots from the perspective of a female Chihuahua mix, an unlikely partner in the deep forests, yet equipped with an ability to "read" the landscape through her own sensorial and phenomenological traits; microscopes reveal the textures of feces from different species; images fade into one another; infrared night vision captures the eyeshine of nocturnal creatures; time-lapse accelerates the successive passage of humans, nonhuman animals, cars and motorbikes along the same trails; long, atmospheric shots absorb (non)human subjects in the sensuousness of the wilderness and capture the humidity, fog and temperatures of the weather; images focused on insects, leaves, moss and dew refuse to turn away from the world of the small; there are recordings of rotoscoping projected in natural spaces; and GPS units mark the migratory trail of wolves. Transitions between sequences are rarely smooth. Instead, montage is arranged as sequential cuts between types of shots, cameras and cinematographic styles, drawing the viewer's attention to the interstices between forms of digital representation.

The visual pastiche fleshes out a digital geography that does not offer the twenty-first-century media consumer a site for serenity and repose. The wilderness and natural landscapes are not imbued with romanticized aesthetics of pristineness and purity, because the wilderness of the Anthropocene can no longer be represented in those terms. The notion of "wild" is subjected to a crisis of visual indexing. In the Anthropocene, the "wild" is besieged, destroyed, annihilated and heterogeneous to human presence. As such, a "truthful translation" of the wild into media form will necessarily involve self-reflexive types of cinematography that emerge as tokens of human artificiality. In this manner, *Salvaxe, salvaxe* produces topographies of the uncanny – the unsettling pictures of a wilderness of and in time whose image is revisited from a place in history defined by its reliance on digital technology as an artisan of (post)truth.

Viewers' everyday exposure to the sounds and textures of digital technology does not – paradoxically – increase their sense of familiarity with the wolf. The film defies viewers' capacity to know what to make of the nonhuman subject. The marriage between self-signifying digital representation and the wilderness embodied by the wolves – at times peculiar, at times eerie and perhaps even aberrant – is indicative of the limits of human epistemologies when it comes to confronting species otherness. In the Anthropocene, this is an understanding that we must come to terms with: as Fonseca pushes wolves further into the spectral, the film convincingly suggests that theirs is a presence that can only be ethically encountered through image representation. The "domestication" of wolves through digital technology, therefore, is not meant to upgrade or secure scopophilic desire for possession. If anything, the sense of uncanniness functions as a means to "re-wilden" the spectatorial gaze, nurturing an ontological distance between humanness and wolf-ness, yet fostering a theme of interspecies connectivity based on the sharing of space and time. The film suggests, rather successfully, that this is what the post-human interspecies encounter should look like.

Soundscapes reinforce the Anthropocene's dislocation of the interspecies encounter and create an immersive (and at the same time dissonant) experience. Field recordings constitute the raw material that is digitally treated to subvert anthropocentric expectations of the "sound of the wild." Sound artist Xoán-Xil López uses *in situ* captures and absence recordings, along with captures of infra- and ultrasounds to reveal the sonic universe that animates a world beyond human perception (*Salvaxe, salvaxe* press kit). Complementing sound design and engineering is a soundtrack that synthetically builds upon the Earth's vibrational, electromagnetic spectrum. The film's synthetic and field-recorded material resonate in soundscapes where one feels that frequency, vibration, volume, timbre and texture are compressed into a distorted matter that alienates our sensorial apprehension of the species other. This alienation or de-familiarization, the film would suggest, is the sonic site from which to reeducate and reconstruct interspecies relations and encounters that are consistent with a post-postmodernist ethical response.

This is the Anthropocene at its digital finest: the portrayal of the "glocal" as a territory haunted by specters of fading worlds. The disappearance of wolves and other wildlife goes hand in hand with the extinction of nonhuman ways of phenomenologically experiencing the world. In this light, Fonseca's auteurism reflects what Smaill describes as "the capacity of the camera to function on biopolitical terms, not simply recording but also organizing life for the viewer" (2016, 128). *Salvaxe, salvaxe*'s biopolitical statement draws on the spectrality of wolves, whom humans can only visually attempt to apprehend (but never truly possess) through technological mediation. As one of the captions states over a black screen:

> Even if we wish for their presence and look for them,
> a wolf makes for a difficult image.
> Almost too easy to look at,
> it does not show the chasing and violence,
> the hundreds of photos wolves are not in,
> all the places they are absent from,
> the huge difficulty of obtaining the image of a ghost.
>
> Watching others in the Anthropocene gets tricky.[1]

What does it mean that a wolf is "too easy to look at"? Fonseca challenges the simplicity of the imagery of mythical and charismatic megafauna, implying that the negative spaces and absences that compose the image of a wolf are cryptograms of a history of extermination that has left its residual, ghostly inhabitants suspended in a digital space of disappearance. *Salvaxe, salvaxe*, in this sense and to a certain extent, complies with what María Teresa Lajoinie Domínguez (2024) describes as the spectral archive of the migratory nonhuman peoples that are at the center of wildlife films. As beings that instinctively avoid human encounters, the wolf-specters – in their state of always disappearing – are digitally captured in forest crossroads and paths that the text-based narration calls "asynchronous meeting places." These photographic captures bear the promise of an archive; yet the archival image, as one of the many pieces that are recollected and meaningfully rearranged for the preservation of memory (Alzate Giraldo et al. 2021, 76), makes it inherently "difficult" to look at a wolf, and disrupts the epistemological assumptions of archiveology.

In her study of the archival image in cryptozoological television shows, Zoë Anne Laks (2023) builds on Paul Sheehan's (2008) claim that nonhuman animals are "anti-cinema" because of their ontological imposition against the representational grammars of film. Laks borrows this idea to discuss the "unruliness of animal archival imagery" (2023, 72), which foregrounds the facticity of the nonhuman body and its resistance to anthropocentric narratives. Referring to Animal Planet's show *Lost Tapes*, she notes how the unruly archival image counters archiveological forces of re-appropriation and re-activation, and "demonstrates the potential to both remind us of the ideological and material circumstances of an image's production and activation, as well as propose new ways forward for the future of archival praxis" (2023, 75). *Salvaxe, salvaxe*'s self-reflexive compilation of images of wolves echoes this idea: the elusiveness of real wolves, the (ethical) impossibility of unmediated encounters and the film's insistence on their spectrality transform the captured image into a "'counter-archive' which makes space for animal resistance to [anthropocentric] structures," thus allowing the wolves to posit their own "posthuman 'un-archiveability'" (Laks 2023, 78). In this sense, *Salvaxe, salvaxe* intentionally "sabotages" its anthology of cyborg wolves by thwarting their archiveability. Fonseca's grammars of cinematography bring into prominence forms of interspecies and intermedial connections that are antithetical to the anthropocentrism of the archive, and propose new models for the rearrangement and compilation of nonhuman animal imagery.

There are more elements that align Fonseca's trope of the technological, as form, with a post-postmodernist commitment to truth and ethics. The film's resolute struggle against the Anthropocene is largely grounded in the confection of hybrid subjects and epistemologies of which the digital image is an index. Well into the film, the focus shifts to the matter of species purity and the links between early Christian demonization of wolves and current myths and post-truth misinformation about miscegenation, lineage and the conservationist efforts to rehabilitate packs. The captions quote simple-sentenced maxims that continue to antagonize the species and pit them against local human populations. These beliefs are rife with negative affect: "They aren't like they used to be," "[T]heir snouts are different," "[T]hese look bluer," "[T]hey breed and release them now," "[T]hey're not Iberian wolves," "[T]hey're laboratory wolves." Suspicions about undesirable miscegenation with dogs reveal a toxic human obsession with purity. In a personal email, Fonseca described the creative process behind the sequence as an effort to produce an "unnatural" representation that evoked spectrality while simultaneously withdrawing from the mythology and stereotypes of the big bad wolf. To match the textual narration's escalating semantics about blood purity and cleanness, species shape, color and the metaphor of the laboratory as womb, Fonseca creates a phantasmagoric, oneiric and immersive scene by rotoscoping camera trap images of wolves, projecting them on natural forest spaces and digitally animating blue-hued, blinking textures over their bodies. As they move, these textures eject particle-like flickers that are reminiscent of phosphene, invoking the film's thematic concern with the traces that all creatures leave behind in time and space. The visuals reflect the wolf's embodiment as a crucible of the "monster," the canine and the wild – a creature lived by humans as a discourse. This gives the textual narration the opportunity to define the notion of species as a "changeable concept that cannot contain the complexity of life."

The questioning and blurring of species boundaries go hand in hand with the film's repeated demonstration of (digital) technology as a maker of "mutts." The wolf specter is a historiographic site of human fears and emotions, and its mediatized image reconfigures its body as a cyborg. But wolves are not the only creatures subjected to a cinematographic pastiche that references its own hybridity. The film's sound design experiments with the blending and manipulation of recorded noises and voices of the forests – sometimes using frequencies beyond the limits of human hearing. Scientists and researchers scout the mountains and howl like wolves. The written text replaces and de-authorizes the figure of the voice-over narrator and produces a discourse where scientific registers intersect with poetic wordsmithing. Images of locations where GPS-tracked wolves were killed force the violence of the past and the grief of the present to come together as an act of proper mourning. All these elements circle around the idea of "impure" ways of being and knowing, with digital technology providing the means to represent non-linear and rhizomatic forms of existence – a hybridity built on the (a)synchronous encounters between beings in time and space, where all subjects traverse and leave their trace.

One particularly singular scene shows nighttime camera trap footage of a wolf whose eyeshine, appearing pixelated, shifts form with every still as the shot gradually gets closer to the face. The sequence of stills has a jerking and stuttering effect on the wolf's actual movement, and the eeriness and volume of the sound heighten the alienness of the cyborg-wolf of geometricized eyes. In this way, the subject becomes a discourse and enters the digital network of intertextual relations, for these techniques are reminiscent of the neo-horror genre that, according to Bruno Surace (2019), conceives the camera as a body that is encountered by the monster and forces testimony over narration. In Fonseca's scene, the sequentially enlarged stills, slowly zooming in closer, and the pulse of this digital movement "domesticating" the wolf's motion, reflect the human eye's attempt to make sense of the "lucky shot" (the "found footage") of the creature/monster. Meaning is sought by getting closer and by looking into the eyes of the creature; but the only thing that viewers of *Salvaxe, salvaxe* encounter with this attempt at proximity are the semiotics produced by technology itself – the geometrical whiteness of the digital capture of the eyeshine. The scene exemplifies what Richard Wallace (2021) describes as current mockumentary horror film's play with the impurity of the monster. Digital technologies subvert mockumentary's combination of categorical differences based on the fictional/nonfictional divide, bringing them together "in one place to create impure forms that are both physically and cognitively threatening" (2021, 525). Identifying a signature element of the genre, Wallace refers to shots with "too much texture"; these are characterized by a visual smear caused by the slowing down of sudden movements, or by enlarged frames that reencode monstrosity in indecipherable film grain (2021, 532). In an age marked by the post-truth cultural turn, Wallace argues, mockumentary horror replaces the "big reveal" of the monster with documentary's failure to produce factuality. While it would be grossly inaccurate to claim that Fonseca is pursuing a wolf-monster in the horror-genre sense (nor is the film a mockumentary), the sequence does comply with the premise that digital technology can only close-up on the ultimate unknowability of the creature, making a specter out of it. Camera trapping is a "monstrosity" in itself that smudges and blurs the conventions of space and time as dictated by the rules of the unmediated interspecies encounter. Its deixis is premised on impure epistemologies, and this is as far as the posthuman design will take the viewer. The Anthropocene offers no scenario in which true knowledge about the wolf subject, or the desire for a "natural" encounter with them, can be satisfied.

Salvaxe, salvaxe's self-reflexive trope of technology, however, does not frame these moments as futile or disappointing: instead, the film's persistent motif of impurity, hybridity and "mutt-ness" (the "monstrosity") reclaims the technological encounter with the nonhuman other as a necessity to ensure their survival in the Anthropocene. Asynchronous, digital encounters become the viable solution for ethical engagement. The ever-disappearing wolf-specter's howl urges humans to remain physically absent, to let wild nonhuman others tread the land of the unseen, and to find wonder and satisfaction in technology's magnified appreciation of their unknowability.

A final note should also be said regarding *Salvaxe, salvaxe*'s concern with green shooting. The final credits state that the film was sustainably produced following vegan principles and that all wildlife images were obtained with permission from the appropriate Galician and Portuguese conservation authorities.[2] This information is also emphasized in the press kit along with more extensive observations about the environmental impact of filming in wilderness areas. More importantly, these concerns are also directly addressed in the film. In a scene in which recording devices are being secured on trees, the text reads:

> Technology helps to detect hiding wildlife, reducing human presence in sensitive places, increasing the distances for us to approach that wildlife, revealing what we are not able to see, helping us know without destroying. But technology also fills mountains with batteries and electronic devices to control what by definition must escape human control: wildlife, which flows independently. Whose are the data obtained from wild creatures? Who decides what we can obtain from others?

The retrieval of wildlife records, therefore, leaves a trace of its own that should give us pause. In addition, the documentary's interest in technology faces another major ethical concern in wildlife filming that has famously been discussed by Brett Mills (2010): the violation of nonhuman animals' privacy and the appropriation of their bodies as objects of visual consumption. Every behavior and choice made by wolves indicates a desire not to be seen – a desire that, as stated earlier, *Salvaxe, salvaxe* attempts to satisfy as best as possible by limiting the encounter with the human to the technological mediation of the cyborg-wolf-specter. In the scene where recording devices are being placed, Fonseca more explicitly – and with a refreshing sincerity – steps into his role as a post-postmodernist filmmaker by self-reflexively interrogating the ethics that determine creative choices, suggesting that there are, at best, approximations to more ethical (but not completely innocuous) filmmaking. The avoidance of crittercams (cameras attached to nonhuman animal bodies) and the use of minimally invasive recording methods are important, but there lingers the question of whether it is enough. As Mills states,

> [t]o look at an animal – and to decide that humans have a right to look at animals because animals don't have a right to privacy – is an act of empowerment, reinforcing the moral hierarchy which legitimizes the act in the first place.
>
> (2010, 199)

Salvaxe, salvaxe does not shy away from this dilemma and inherently acknowledges that although trap footage and archival images are less invasive, technology, a tireless producer of armies of specters, still remains an index of human power over other species.

Conclusion

In his recent book *A Natural History of the Future* (2022), biologist Rob Dunn delves into the inescapability of biological, evolutionary and ecological laws and what they mean (and imply about the hubris of humanity) on a planet damaged by anthropogenic wipeout. Dunn discusses the concept of "homing," a term coined by biologist Bernd Heinrich (2014) to refer to the behavior whereby species seek out (or migrate back to) places that meet the conditions for their wellbeing and continuity. Dunn foresees that:

> Most species on Earth will need to move to survive climate change. . . . They will need to move from their present islands of habitat to different islands of habitat where their preferred conditions can newly be found. They will need to move to find their new home.
>
> (2022, 61)

He later adds that:

> For many species, there will be no new home. They will wander and never find what they need. Or won't find it in time. Or they will arrive and the climate will be perfect, but something else will be missing.
>
> (2022, 62)

Dunn's scientifically informed vision paints a somber picture of nomads and homeless travelers attempting to survive on a planet that increasingly struggles to meet their needs and places them on a path toward resourcefulness or extinction. His book collects the findings and efforts of scientists who navigate theoretical and practical counteractions against the Anthropocene and the mishandling and disregard of evolutionary laws. This chapter has examined strategies that documentary filmmakers have devised to carve out narratives of hope in spite of the feelings of doom and despair that the Anthropocene and its discourses have brought forth. In their own way, they form aesthetic roads and pathways that all animals can tread as we follow our homing instincts toward new discourses and audiovisual representations through which to reach an ethical interspecies engagement, based on respect and connectivity.

The films discussed earlier explore in their own ways how the perceived barrenness of the postmodernist aesthetic falls short of illuminating remedial passages leading to hope and perseverance. In line with twenty-first-century post-postmodernism, they construct narratives that suffuse postmodern self-reflexivity with an ethical agenda committed to truth and the "real." *Carnage* salutes the mockumentary genre with its careful dosing of parody the better to wrap the horrific facts of factory farming and carnism in palatable discourses that invite viewers to reflect on the pain, absurdity, disgust and (environmental) injustices caused by animal agriculture. *Racing Extinction* uses the trope of (digital) technology as a thematic device to inspire viewers to act. It challenges archiveology's preoccupation with the past by proposing forms of technological representation of nonhuman others whose beauty compels the beholder to *do something* for critically endangered species, thereby shifting the focus onto the future. *Salvaxe, salvaxe* subjects (digital) technology to a self-reflexivity based on form, confronting the spectrality of other species from a post-postmodernist aesthetic of the liminal that finds in the asynchronous, mediatized encounter the means to re-educate the human gaze in the Anthropocentric age, making it more ethical.

Because, as shown throughout this book, advocacy and activism are actions that are antithetical to moral stagnancy, transformative change is always at the center of animal advocacy nonfiction film. Hope, therefore, manifests itself more or less explicitly in all the films that are part of the subgenre. The colossal damage of the Anthropocene represents the ultimate consequences of ecocide and the institutionalization of speciesism. Filmmakers understand the magnitude of the crisis and that real, effective change must come at a political and institutional level in order to have a sizeable impact. Nonetheless, as evinced in this book, they also appeal to individual citizens of the Global North, holding every consumer accountable, but at the same time empowering them by reminding them of their freedom of choice in everyday decisions pertaining to their lifestyle. Animal advocacy documentary films will continue adding titles to their name as long as the animal-industrial complex persists within capitalist economies, as long as objectification of other species remains normalized, and as long as the (slow) violence inflicted upon nonhuman victims is ignored and morally excused. But mostly, the subgenre will continue to grow as long as the flame of hope keeps on burning.

Notes

1 Quotations are taken from the official English translation of the text-based narration. I also maintain the centered text formatting used in the film to illustrate the original phrasal cadence.
2 The credits list the Dirección Xeral de Patrimonio Natural da Xunta de Galicia and the Instituto da Conservação da Natureza e das Florestas de Portugal (ICNF).

References

Adams, Matthew. 2022. "Communicating Vegan Utopias: The Counterfactual Construction of Human-animal Futures." *Environmental Communication* 16 (1): 125–38. https://doi.org/10.1080/17524032.2021.1969977.
Alber, Jan, and Alice Bell. 2019. "The Importance of Being Earnest Again: Fact and Fiction in Contemporary Narratives Across Media." *European Journal of English Studies* 23 (2): 121–35. https://doi.org/10.1080/13825577.2019.1640414.
Alzate Giraldo, Alejandro, Claudia María Maya Franco, and Verónica Heredia Ruiz. 2021. "El archivo audiovisual y la construcción de la memoria a través del formato documental para television." *Enunciación* 26 (1): 74–90. https://doi.org/10.14483/22486798.16992.
Amstell, Simon, dir. 2017. *Carnage: Swallowing the Past*. BBC.
Andersen, Kip, and Keegan Kuhn, dirs. 2014. *Cowspiracy: The Sustainability Secret*. A. U. M. Films, First Spark Media.
Armiero, Marco. 2021. *Wasteocene. Stories from the Global Dump*. Cambridge University Press.
Armiero, Marco, and Massimo De Angelis. 2017. "Anthropocene: Victims, Narrators, and Revolutionaries." *South Atlantic Quarterly* 116 (2): 345–62. https://doi.org/10.1215/00382876-3829445.
Berger, John. (1980) 2009. *Why Look at Animals*? Penguin.
Brockway, Otto, and Ludovic Brockway, dirs. 2021. *Eating Our Way to Extinction*. Broxstar.
Canty, Kristin, dir. 2011. *Farmageddon*. Gravitas Ventures.
Chang, Chia-Ju. 2017. "Global Animal Capital and Animal Garbage: Documentary Redemption and Hope." *Journal of Chinese Cinema* 11 (1): 96–114. https://doi.org/10.1080/17508061.2016.1269483.
Crutzen, Paul J., and Eugene F. Stoermer. 2000. "The 'Anthropocene.'" *IGBP Newsletter* 41: 17–18.
Davidson, Kief, and Richard Ladkani, dirs. 2016. *The Ivory Game*. Terra Mater Studios, Netflix, Vulcan Productions.
Dunn, Rob. 2022. *A Natural History of the Future*. Basic Books.
Duvall, John A. 2017. *The Environmental Documentary: Cinema Activism in the 21st Century*. Bloomsbury.
Ehrlich, Pippa, and James Reed, dirs. 2020. *My Octopus Teacher*. Off the Fence, The Sea Project, Netflix.
Farkas, Johan, and Jannick Schou. 2020. *Post-Truth, Fake News and Democracy. Mapping the Politics of Falsehood*. Routledge.
Fonseca, Emilio, dir. 2024. *Salvaxe, salvaxe*. Walkie Talkie Films.
Fulkerson, Lee, dir. 2011. *Forks Over Knives*. Monica Beach Media.
Guggenheim, Davis, dir. 2006. *An Inconvenient Truth*. Participant Productions, Paramount Classics, Lawrence Bender/Laurie David.
Heinrich, Bernd. 2014. *The Homing Instinct. Meaning and Mystery in Animal Migration*. Houghton Mifflin Hartcourt.
Horak, Jan-Christopher. 2006. "Wildlife Documentaries: From Classical Forms to Reality TV." *Film History: An International Journal* 18 (4): 459–75. https://doi.org/10.2979/fil.2006.18.4.459.
Hutcheon, Linda. 1988. *A Poetics of Postmodernism: History, Theory, Fiction*. Routledge.
Hutcheon, Linda. (1989) 2002. *The Politics of Postmodernism*. Routledge.

Jameson, Fredric. 1991. *Postmodernism, or The Cultural Logic of Late Capitalism*. Duke University Press.
Kenner, Robert, dir. 2008. *Food, Inc*. Magnolia Pictures, Participant Media, River Road Entertainment.
Kolbert, Elizabeth. (2014) 2024. *The Sixth Extinction. An Unnatural History*. Holt.
Lajoinie Domínguez, María Teresa. 2024. "*Le peuple migrateur* (2001) o el documental de vida salvaje como archivo espectral." *IC – Revista de Información y Comunicación* 21: 95–115. https://doi.org/10.12795/IC.2024.I21.05.
Laks, Zoë Anne. 2023. "Excavating Animal Planet's *Lost Tapes*: The Unruly Images of the Posthuman Counter-Archive." *Canadian Journal of Film Studies* 32 (1): 61–86. https://doi.org/10.3138/cjfs-2021-0040.
Lippit, Akira Mizuta. 2000. *Electric Animal. Towards a Rhetoric of Wildlife*. University of Minnesota Press.
Marchessault, Janine. 2017. *Ecstatic Worlds: Media, Utopias, Ecologies*. MIT Press.
McKibben, Bill. (1992) 2006. *The Age of Missing Information*. Random House.
McLaughlin, Robert L. 2012. "Post-Postmodernism." In *The Routledge Companion to Experimental Literature*, edited by Joe Bray, Alison Gibbons, and Brian McHale, 212–23. Routledge.
Mills, Brett. 2010. "Television Wildlife Documentaries and Animals' Right to Privacy." *Continuum. Journal of Media & Cultural Studies* 24 (2): 193–202. https://doi.org/10.1080/10304310903362726.
Nixon, Rob. 2011. *Slow Violence and the Environmentalism of the Poor*. Harvard University Press.
Nussbaum, Martha C. 2022. *Justice for Animals. Our Collective Responsibility*. Simon & Schuster.
Petersen, Christer. 2023. "Stranger than Fiction: On Alternative Facts and Fictional Epistemologies." In *Truth and Fake in the Post-Factual Digital Age*, edited by Peter Klimczak and Thomas Zoglauer, 73–88. Springer.
Pierschel, Marc, dir. 2017. *The End of Meat*. Blackrabbit Images.
Plantinga, Carl. 2009. *Moving Viewers. American Film and the Spectator's Experience*. University of California Press.
Psihoyos, Louie, dir. 2009. *The Cove*. Ocean Preservation Society, Diamond Docs, Skyfish Films.
Psihoyos, Louie, dir. 2015. *Racing Extinction*. Okeanos, Discovery, Oceanic Preservation Society.
Quinn, Christopher, dir. 2017. *Eating Animals*. Handsomecharlie Films, Heretic Films, Big Star Pictures.
Rangel-Buitrago, Nelson, William Neal, and Allan Williams. 2022. "The Plasticene: Time and Rocks." *Marine Pollution Bulletin* 185 (B): 114358. https://doi.org/10.1016/j.marpolbul.2022.114358
Rose, Deborah Bird. 2011. *Wild Dog Dreaming. Love and Extinction*. University of Virginia Press.
Salvaxe, salvaxe press kit (English). 2024. https://salvaxesalvaxe.walkietalkiefilms.com.
Sánchez-Vilela, Rosario. 2019. "Falso documental y series de television. El género y sus vertientes en la ficción televisiva uruguaya." *Comunicación y Medios* 28 (40): 56–67. http://dx.doi.org/10.5354/0719-1529.2019.52919.
Sartore, Joel. 2015. "Projecting Change: The Empire State Building." *Joel Sartore News*, August 3. https://www.joelsartore.com/projecting-change-the-empire-state-building.
Sauper, Hubert, dir. 2004. *Darwin's Nightmare*. Mille et Une Productions, Coop99, Saga Film.
Sen, Shaunak, dir. 2022. *All That Breathes*. HBO, Rise Films, Submarine Deluxe, Sideshow, HHMI Tangled Bank Studios.
Shaw, Meghan, Matilda Dunn, Sarah Crowley, Nisha Owen, and Diogo Veríssimo. 2024. "Using Photo Editing to Understand the Impact of Species Aesthetics on Support for Conservation." *People and Nature* 6 (2): 660–75. https://doi.org/10.1002/pan3.10602.
Sheehan, Paul. 2008. "Against the Image: Herzog and the Troubling Politics of the Screen Animal." *SubStance* 37 (3): 117–36. https://doi.org/10.1353/sub.0.0023.
Shukin, Nicole. 2009. *Animal Capital. Rendering Life in Biopolitical Times*. University of Minnesota Press.

Sklair, Leslie. 2021. "Editor's Introduction." In *The Anthropocene in Global Media: Neutralizing the Risk*, edited by Leslie Sklair. Routledge. https://iiraorg.com/the-anthropocene-in-global-media.
Smaill, Belinda. 2016. *Regarding Life. Animals and the Documentary Moving Image*. State University of New York Press.
Surace, Bruno. 2019. "The Flesh of the Film: The Camera as a Body in Neo-Horror Mockumentary and Beyond." *Northern Lights: Film & Media Studies Yearbook* 17 (1): 25–41. https://doi.org/10.1386/nl_00003_1.
Tabrizi, Ali, dir. 2021. *Seaspiracy*. A. U. M. Films, Disruptive Studios.
Thompson, Monte, dir. 2010. *Call of Life: Facing the Mass Extinction*. Species Alliance.
Thunberg, Greta. 2019. *Speaking Notes for a Speech Held in the European Parliament at Strasbourg at an Extraordinary Meeting of the Environment Committee, Open to All Members*. European Parliament, April 16. https://www.europarl.europa.eu/resources/library/media/20190416RES41665/20190416RES41665.pdf.
Vermeulen, Timotheus, and Robin van den Akker. 2010. "Notes on Metamodernism." *Journal of Aesthetics & Culture* 2 (1): 2–14. https://doi.org/10.3402/jac.v2i0.5677.
von Einsiedel, Orlando. 2014. *Virunga*. Bertha Foundation, Britdoc Foundation, Grain Media, Violet Films, Appian Way Productions.
Wallace, Richard. 2021. "Documentary Style as Post-Truth Monstrosity in Mockumentary Horror Film." *Quarterly Review of Film and Video* 38 (6): 519–40. https://doi.org/10.1080/10509208.2020.1780107.
Weik von Mossner, Alexa. 2021. "Screening Veganism: The Production, Rhetoric, and Reception of Vegan Advocacy Films." In *The Routledge Handbook of Vegan Studies*, edited by Laura Wright, 309–18. Routledge.
Weik von Mossner, Alexa. 2025. "Envisioning Intergenerational Justice: Hope, Despair and Transformative Action in Climate Change Films." In *Cinema of/for the Anthropocene*, edited by Katarzyna Paszkiewicz and Andrea Ruthven, 47–62. Routledge.

Index

For Product Safety Concerns and Information please contact our EU
representative GPSR@taylorandfrancis.com
Taylor & Francis Verlag GmbH, Kaufingerstraße 24, 80331 München, Germany

www.ingramcontent.com/pod-product-compliance
Ingram Content Group UK Ltd.
Pitfield, Milton Keynes, MK11 3LW, UK
UKHW020139200926
475702UK00005B/820

* 9 7 8 1 0 3 2 6 6 1 3 5 3 *